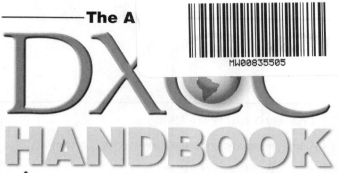

The ARRL

DXCC
HANDBOOK

*Worldwide ham radio operating
and the ARRL DXCC Award*

By Jim Kearman, KR1S

Production
Michelle Bloom, WB1ENT
Sue Fagan, KB1OKW—Front Cover
Jodi Morin, KA1JPA
David Pingree, N1NAS

Table of Contents

About the Author

Jim Kearman, KR1S, has been licensed since 1962. He discovered DXing shortly after receiving his General class license, and the call sign WB2EDW, in 1963. After several years of inactivity punctuated by military service and life in general, he received his first DXCC award in 1991. Jim now specializes in 80 and 40 meter operation from a condo in Stuart, Florida, using temporary wire antennas. He's been a Life Member of ARRL since 1976, and is also a member of the Old Old Timers Club (OOTC). Jim's Web site, **kr1s.kearman.com/**, contains several links of interest to DXers; his e-mail address is **kr1s@kearman.com**.

Acknowledgments

For encouragement and suggestions, my thanks go to Jim Cain, K1TN, Larry Keith, KQ4BY, and Warren Stankiewicz, NF1J. Web sites hosted by Rodney Dinkins, AC6V, and Ted Melinosky, K1BV, were invaluable resources. Thanks also to the thousands of DXers, DXpeditioners, QSL managers and the unsung heroes and heroines of the QSL bureaus around the world, who contribute to making DXing such a challenging and rewarding pastime. Finally, thanks to the ARRL Production Department for transforming my manuscript into the book you now hold in your hands.

Stuart, FL
November 2006

Station Resources

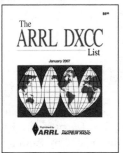

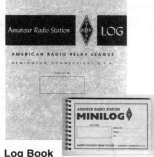

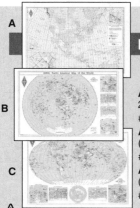

Introduction

Why DX?

B efore you read this book, you have a right to know something about the author. Is this another book by a rich ham with tens of thousands of dollars invested in high-tech gear and tall towers? No!

I've never had a "big-gun" station, although I've been privileged to operate from a few. Experience has taught me that you can have a great time chasing and working distant — and sometimes rare — stations, without investing huge sums of money, or owning acres of land.

In fact, I think my average stations gave me an advantage, because I had to try harder sometimes, and learn techniques the owner of a superstation doesn't need. These techniques — tricks, if you will — have snagged contacts from under the noses of many better-equipped stations. You expect to do well at a big station. True, succeeding on a limited budget requires more from you, the operator, but also pays enormous dividends.

Something about humans makes us competitive, but we have the choice of how much time and treasure to spend on our pursuits. Amateur Radio operators have always sought to lengthen the distances over which they could communicate. It's a way of determining how well our stations — and we, the operators — perform.

At heart, all competitions are battles with ourselves. How well can we do, and how much better? In some sports you can't win without spending a fortune, and even then, success isn't guaranteed. Fortunately, Amateur Radio is different. In Amateur Radio, the key ingredient is the operator. With practice and a little coaching, we can succeed, even excel, with modest equipment and antennas.

So, a key factor in DXing is competitiveness. You don't have to smash the competition, but you have to get your signal out a little farther, hear weaker stations a little better, and learn good operating habits. It's worth noting here that these techniques and habits will serve you well forever, even if you someday graduate to the level of *Big-Gun DXer*. Hitting the target is what matters, not making a lot of noise.

I started DXing in the 1960s, when the Cold War was in full swing. As a high-school student well-versed in the sins of communism, it was exciting to work stations behind the Iron Curtain. In those days, Soviet amateurs rarely gave more than basic information during a contact, but you couldn't escape the feeling that on the other end of the QSO sat a ham much like yourself. Working stations in many countries can't help but change your worldview. People in foreign countries may not be just like us, but they have ham radio, so how different can they be?

Amateur Radio is a global collective of men and women with similar interests. When you start communicating outside your own country, you put aside some of your prejudices, and become a citizen of the world. What a wonderful thing it is to talk directly with someone in another country, without the intermediaries of Internet service providers or telephone companies, and essentially without censorship.

Some hams are put off by DXing because many contacts are short, and limited by language differences. As Marshall McLuhan wrote, though, "the medium is the message." Many of us now engage in online discussions across national borders, but Amateur Radio is the next best thing to talking with someone face-to-face. In these troubled times we too often focus on the differences between nations. DXers seek commonality, lessening differences while emphasizing our common interest in radio communication. What's more, even the simplest DX contacts require more skill than simply knowing how to plug in and log on. You can't buy DX QSOs, you have to earn them.

Something else that's always excited me about DXing is the organic medium we rely on, the ionosphere. Run some wires from Point A to Point B and you expect to communicate. Not so for the DXer. The ionosphere is anything but predictable. Knowing when to look for DX on a particular band is helpful, but I've also done well by being open to unexpected propagation paths. Not to sound New Age-ish, but DXing gets you close to Nature.

Unlike broadcast stations, DX stations don't keep regular schedules. Uncertainty, coupled with the occasional unusual band opening, puts Amateur Radio communication in a class of its own. As licensed radio amateurs we have the privilege of exploring these amazing phenomena. I have an Internet connection and a cell phone, but they haven't diminished my pleasure in tuning across the ham bands. As well as being more or less competitive, DXers haven't been jaded by modern technology.

Not that DXers are diehards clinging to obsolete technology. Advances in communications science have come to Amateur Radio as well. The radios of today are far advanced over those of my youth. Personal computers and the Internet have changed Amateur Radio as much as any other area of our lives. Yet, even with all the benefits of modern science, DXing still comes down to individuals far apart, trying to hear and communicate with one other. There is still a place for the individual, still room for a person to excel by diligence, that is missing in the realm of modern communications media.

DXing allows me to participate at a deeper level; and that pays dividends in feelings of accomplishment. Long ago I was seduced by the mysterious sound of distant signals, bouncing off the ionosphere and fighting their way through the auroral curtains. After more than 40 years, those fluttery signals from faraway places still thrill me. In short, I chase DX because it's fun!

I hope this book will inspire you to try DXing, and that you'll use it as a guide to beginning your own DX journey. Let's go!

Chapter 1

DXing and the
DXCC Award

What is DXing?

DX ing is the art and science of communicating with distant stations, usually in foreign countries. DXing on the UHF and microwave bands may involve terrestrial contacts spanning only a few hundred miles, and many of the operating techniques presented in this book will be useful to radio amateurs operating on those frequencies. As this book is devoted to radio amateurs working for the ARRL DX Century Club (DXCC) award, we'll think in terms of countries, and more specifically, DXCC *entities*.

The basic DXCC award requires you to prove you have had two-way contacts, on Amateur Radio frequencies, with at least 100 DXCC entities, not all of which are technically countries in their own right. More on this later. First, let's take a quick tour of DXing history.

History of DXing

Guglielmo Marconi sent signals across the Atlantic Ocean in 1901, but radio amateurs did not succeed in "crossing the pond" for more than another 20 years. Transatlantic tests sponsored by ARRL and the International Amateur Radio Union (IARU) inspired operators on both sides to transmit and listen, and the first successful two-way contact finally happened in November 1923. DXing was born!

If you can, read the original *QST* article.[1] Spark transmitters were still in use then, but stations running vacuum-tube CW transmitters had the best results. The best receivers of the time were a far cry from those of today, yet some spark signals did make it across the Atlantic. The frequencies used were in what is now the AM Broadcast Band, about 1500 kHz. Some amateurs think they'd need high power and big antennas to work DX, so how did those old timers do it with such limited equipment? Simple: No one told them they couldn't!

Until cross-border and intercontinental Amateur Radio communication became commonplace, licensing authorities gave no consideration to country-specific prefixes. It was likely that the call sign 1AA could be assigned in a half-dozen or more countries. For a few confusing years, many radio amateurs made up their own prefixes. The situation finally got sorted out, and the first call sign prefixes were assigned.

What is DXCC?

Beginning in 1932, ARRL began compiling a list of countries, to serve as a standard for DXers. The DXCC award did not yet exist, but avid DXers were already keeping score. In a 1935 *QST* article, Clinton B. DeSoto, W1CBD, laid out many of the criteria still used to determine which entities count for DXCC credit.[2] To qualify for DXCC you must prove you had two-way communications on Amateur Radio frequencies, with at least one station in at least 100 different countries or DXCC entities (see below) listed on the *DXCC Countries List*. A contact with another Amateur Radio station in your own country counts for credit. This chapter summarizes many DXCC Rules, which are included in Appendix A in this book. The Rules often change, however, so check the online Rules page at **www.arrl.org/awards/dxcc/rules.html** for the latest version.

DXCC Entities

DeSoto wrote, *"Each discrete geographical or political entity is considered to be a country."* Of the two, the political country, now called an *entity*, is easiest to understand. To be considered in this category, a political entity must be:

- A member state of the United Nations
- Have been assigned a call sign prefix block by the International Telecommunications Union (ITU)
- Have a permanent population

The last several years have seen many changes in Political entities, especially in Europe. Before the collapse of the Soviet Union, there were two Germanys, whereas several Political entities of today once constituted the single entity of Yugoslavia. If you confirmed contacts with East Germany and Yugoslavia before they restructured themselves, you can count those contacts for DXCC, with one exception, noted later in the discussion of DXCC awards. Entities that once counted for DXCC credit, but no longer do, are called "deleted entities."

Geographical entities are more complicated. For example, the states of Alaska and Hawaii count as separate DXCC entities. Alaska qualifies because it is separated from the contiguous 48 states by more than 100 km (61 miles), by another country, Canada. Hawaii counts because its islands are separated from the contiguous 48 states by more than 350 km (214 miles), and no other islands that are part of the United States lie in between, shortening the distance.

There is a third category of DXCC entities, called *Special Areas*, which covers anomalies like the club station at ITU Headquarters in Geneva, Switzerland, the club station at United Nations Headquarters in New York City, and other entities not covered in the Political or Geographical categories.

Just as new entities are added to the DXCC List, some entities are deleted when they no longer meet the criteria for inclusion. Examples include the former countries of East Germany and West Germany. After reunification, those two countries were deleted, and a new country, the Federal Republic of Germany, was added. If you worked East Germany and West Germany before reunification, your DXCC total reflects those countries, but they do not count toward the DXCC Honor Roll or DXCC Challenge (see Chapter 11).

DXCC Awards, a Brief Summary

There are several DXCC awards. Chapter 11 covers them in detail. Most DXers begin with the Mixed award, which allows contacts on any legal mode, on any amateur frequency, except contacts made through repeater stations or amateur satellites. There is a Satellite DXCC, however, but contacts made through other forms of repeater stations do not qualify for any DXCC award.

If you specialize in voice, radio-teletype (RTTY) or Morse code (CW) operation, there are separate DXCC awards for those modes. You can also apply for band-specific awards, for any band from 160 to 2 meters. Eventually, you will want to apply for Five-Band DXCC (5BDXCC), for confirmed contacts with 100 current DXCC entities (deleted entities don't count for 5BDXCC) on the 80, 40, 20, 15 and 10 meter bands (using any legal mode). 5BDXCC can be endorsed for the other bands, from 160 to 2 meters, and a beautiful plaque is available as well.

The top rung of the DXCC ladder is the *Honor Roll*. To qualify, you must have a total confirmed entity count that places you among the numerical top ten DXCC entities total on the current DXCC List. For example, if there are 337 current DXCC entities, you must have at least 328 entities confirmed. To establish the number of DXCC entity credits needed to qualify for the Honor Roll, the maximum possible number of current entities available for credit is published monthly on the ARRL DXCC Web Page, **www.arrl.org/awards/dxcc/**.

There is DX, and There is DX

While we can agree that a DX station is in a different country from our own, not all DX is the same. DX stations fall into two classes, common and rare, and there are levels of rarity as well. US radio amateurs will, unfortunately, never be considered rare. Rarity is relative, though, and much depends on the band, mode and propagation conditions. Many entities that are fairly common on frequencies higher than 3.5 MHz are scarcely found on the 160, 6 or 2 meter bands.

Most of the time you will be *working* (contacting) DX stations owned and operated by residents of the country or entity. Some countries have very active

Amateur Radio populations. They're easy to work and confirm. Some smaller countries may have only one active radio amateur, or only a few. But even larger countries aren't always well represented on the bands. At this writing, China is a good example. Though the most populous country in the world, China has a relatively small radio amateur population.

Radio amateurs visiting a country often get permission to operate. The lengths of their visits may range from short vacations to several years. Vacationers often use marginal antennas and are harder to work; long-term residents may operate only occasionally.

Then there are those uninhabited islands and reefs, with mysterious (to DXers, anyway) names like Bouvet and Spratly. Unless they were recently activated by a large group of operators, for at least several days, they are rare on any band or mode, regardless of propagation conditions. Don't worry too much about them just yet. Almost every entity comes up eventually, and there are plenty of less-rare entities to work.

What Counts As a DX Contact?

The content of a valid DX contact, or QSO, is sometimes controversial. Except in Amateur Radio contests, where a specific exchange of information is required, a valid contact consists of both stations copying each other's call signs, and some item of information, which may be nothing more than an abbreviated signal report. For example, on CW, both stations may send "5NN," short for 599, or simply "59" on voice, even though neither station registers extremely strong on the other's S meter.

Such brevity turns off many radio amateurs. They feel a contact should include some personal information, at least an exchange of names and cities. There are practical reasons for keeping it short, however, especially when conditions are marginal and the DX station is considered rare. As I mentioned in the Introduction, "the medium is the message." Though the contact may be short, when we contact another station anywhere, we are doing something unique to Amateur Radio. Communicating by Amateur Radio is an accomplishment, which increases with the difficulty of making the contact. People who go to rare DXCC entities usually want to contact as many other stations as possible. If you'd like to chat with them, look them up at a convention sometime, and have an *eyeball QSO*.

On the other hand, if the DX station wants to chat, it is impolite to reply with a signal report and "73." You don't want to hog the frequency, but the other station may appreciate knowing your name and where you're located. There's no need to get more involved than that, unless the DX station asks for more information.

Is DXing Hard?

DXing is a competitive sport. Like any sport, what you get back is proportional to what you put in. The most-important part of any DXer's station is the *operator*. Working your first 100 entities is pretty easy, as there are usually 200 or more on the air at any time. A few contacts will take some effort, but they're the ones you'll value most. Looking back over more than 40 years of chasing DX, the contacts I recall weren't all with rare countries. A few that were, weren't that hard, thanks to the experience I had built up. In time, you will have a store of DX memories, too. You may forget what rig or antenna you had, but you'll savor the satisfaction of working a new one. DXing *is* harder than general operating, but even without the QSL cards and the DXCC certificate, I think the rewards are worth the effort.

What Mode Should I Use?

I'm glad you asked! That means you're flexible enough to use two or more modes. The answer is to operate on every mode your station can handle. You may find you favor one mode over the others, but try to improve your skills in all of them. Much depends on the capabilities of your station. As a general rule, the narrower the bandwidth, the easier it is to communicate. For a given power level you will have more chances to get through on CW than on SSB or RTTY. But what if the entity you want to work is only operating SSB or RTTY?

What Equipment Do I Need?

We often see photographs of highly competitive DX and contest stations. While impressive, they are a small fraction of the stations you'll be competing with. In fact, you probably won't hear the *big guns* very often, except in contests. Most of them have worked all the DXCC entities, on all bands. Pity them, there's nothing left to do. Many, many people have made DXCC while running 5 W or less into average antennas. If you have the choice, a basic 100 W transceiver and the best antennas you can install will more than suffice. What matters is getting on the air and gaining experience. That's something no amount of money can buy. DXing is only a hobby, so spend only what you can afford and learn to make the best of it.

Where Do I Find DX Stations?

You can work DX on any band that has propagation to other countries. Propagation deserves a chapter of its own. I've saved that topic for Chapter 7, because I want to talk about other subjects first. While ARRL has published band plans (online at **www.arrl.org/FandES/field/regulations/bandplan.**

html), less formal band plans are in use. Rod Dinkins, AC6V, has created a Web site containing a staggering amount of information of interest to radio amateurs (**www.ac6v.com/**), from which **Table 1-1** (on the next page) is excerpted, with permission. As used in Table 1-1, a window is a range of frequencies.

Note that frequency and mode allocations are not the same all over the world. Make sure it's legal for you to transmit on a frequency before answering a DX station. (At this point it's worth mentioning that US DXers will benefit greatly from having an Amateur Extra class license. You can achieve DXCC with any class of license, but being able to operate on more frequencies is a major plus. The cost of a license manual or two could be the best investment you can make.)

What's All That Commotion?

You're tuning across the band and suddenly hear station after station, spread out over tens of kHz, frantically sending their call signs. Yes, it's a *pileup*, and somewhere in there is a DX station *running* that pileup. Pileups sound chaotic, but there is a system to them. The trick to breaking a pileup is to figure out the system. The DX station can't be listening on all those frequencies at once. Your task is to learn where the DX station is *going to listen next*, and be there, sending your call, at that moment. I've devoted a complete chapter to just that subject. For now, spend a few minutes listening to the pileup. Do you hear the stations that never change frequency, that keep sending their call signs over and over?

You don't want to be one of them, especially if you want to get your own DXCC certificate. Learning to properly work through a pileup puts you way ahead of DXers who don't know how to, even if they have stronger signals. Unskilled operators can drive you to distraction sometimes. Using the techniques covered in this book, and others you will learn by experience, you'll find yourself spending less time in pileups, and less time being aggravated.

What's A DXpedition?

One term frequently used by DXers is *DXpedition*. DXpeditions take many forms, from casual operations by vacationing radio amateurs, to large-scale operations from rare DX entities, often difficult-to-reach islands or countries from which Amateur Radio signals are rarely heard. These valiant DXpeditions often present the only opportunity you'll ever have to contact these entities. *YASME—The Danny Weil and Colvin Radio Expeditions*, written by James D. Cain, K1TN, and published by ARRL, is an exciting introduction to DXpeditioning, a must-read for every DXer.

DX Nets and Lists

Some DX stations aren't comfortable calling CQ and handling a pileup of stations wanting to work them. Very often, these stations are somewhat rare. DX nets and lists may be helpful in these situations. Most DX nets and lists are limited to SSB operation, and most of them are on the 20 meter band. You shouldn't need to work a station on a net or list to get your first 100 DXCC entities, but eventually something you can't work any other way will show up on one.

TABLE 1-1

DX Calling and Working Frequencies, 160-6 Meters

160 METERS

1800 – 1810 kHz	Some countries cannot operate in this segment
1810 – 1830 kHz	Some other countries cannot operate in this segment
1828.5 kHz	CW DXpeditions frequently here
1830 – 1840 kHz	CW, RTTY and other narrowband modes, intercontinental QSOs only
1840 – 1850 kHz	CW, SSB, SSTV and other wideband modes, intercontinental QSOs only

80/75 METERS

3500 – 3510 kHz	CW DX window
3505 kHz	CW DXpeditions frequently here
3590 kHz	RTTY DX
3790-3800 kHz	SSB DX window
3799 kHz	SSB DXpeditions frequently here

40 METERS

7000 – 7010 kHz	CW DX window
7005 kHz	CW DXpeditions CW frequently here
7040 kHz	RTTY DX
7065 kHz	DXpedition SSB (Listening for USA on 7150 kHz and above)

30 METERS

10.110 MHz	CW DXpeditions frequently here

20 METERS

14.025 MHz	CW DXpeditions, usually listening about 14.025 MHz
14.080 MHz	RTTY DXpeditions
14.100 MHz	Propagation beacons (see Chapter 7)
14.195 MHz	SSB rare DX and DXpeditions transmit here, listening above this frequency

17 METERS

18.075 MHz	CW DXpeditions, usually listening above this frequency
18.110 MHz	Propagation beacons
18.145 MHz	SSB DXpeditions, usually listening above this frequency

Nets and lists are run by stations whose signal is loud enough to be heard by both the DX station and the stations that want to work them. The net-control station (NCS) is often called the *emcee*, for master of ceremonies, though they may not appreciate it if you call them that. When the net control station stands by for new check-ins, you send your call and wait to be acknowledged. Then you tell the NCS which DX stations on the net or list you want to work.

When your turn comes, the NCS calls you. You call the DX station and give a signal report. The DX station replies with a signal report and

---

15 METERS

21.025 MHz	CW rare DX and DXpeditions transmit here, listening above this frequency
21.080 MHz	RTTY DXpeditions
21.150 MHz	Propagation beacons
21.295 MHz	SSB rare DX and DXpeditions, usually listening above this frequency

12 METERS

24.895 MHz	CW rare DX and DXpeditions, usually listening above this frequency
24.930 MHz	Propagation beacons
24.945 MHz	SSB rare DX and DXpeditions, usually listening above this frequency

10 METERS

28.025 MHz	CW rare DX and DXpeditions, usually listening above 28.080 MHz RTTY, rare DX and DXpeditions usually listening above this frequency
28.120 – 28.300 MHz	Propagation beacons
28.495 MHz	SSB rare DX and DXpeditions, usually listening above this frequency

6 METERS

50.06 – 50.09 MHz	Propagation beacons
50.090 MHz	CW calling frequency
50.70 MHz	RTTY Calling Frequency
50.100 – 50.130 MHz	SSB DX window
50.110 MHz	SSB DX calling frequency (usually non-USA Stations call here)
50.115 MHz	DXpeditions frequently operate CW and SSB here

acknowledges receiving yours. Then you acknowledge the DX station's signal report. If the NCS determines that you and the DX station got each other's signal reports correct, you'll hear "Good contact," and you've got a new entity in the log.

This may not sound like a great deal of fun, and often it isn't. My suggestion is to avoid DX nets and lists until there is nothing else on the air you can work for a new one. I think it's better to spend your time searching for DX, and honing your operating skills.

Notes

[1]Also recommended is DeSoto, *Two Hundred Meters and Down*, (ARRL: Newington, 1936) and still in print. This history of Amateur Radio in the United States through 1936 proves the adage, "The more things change, the more they stay the same."

[2]DeSoto, "How to Count Countries Worked," *QST*, Oct 1935, p 40.

Chapter 2

Listen!

What's that sound?

Listening is an art form, in face-to-face conversation or in DXing. Someone is saying something. Are you paying attention? Careful, attentive listening is key to working DX before it shows up on the spotting networks.

A few months ago I was testing a new antenna and homebrew tuner. I was listening on 40 meter CW about an hour before sunset. One European signal was coming through, but it was weak. Tuning across the band I heard a stronger signal, with noticeable fluttering. The station seemed to be ragchewing. The band was open to South America, and someone there was calling CQ a few hundred hertz lower in frequency. But this fluttery signal got me wondering, so I turned on my amplifier, fortunately already tuned up on 40. The signal was so strong it could have been local. But there was that flutter…

Meanwhile, the South American was still calling CQ. I think he was hearing the other station, too, because he sounded pretty upset that someone would be ragchewing so close to *his* frequency. Then the other station signed off. Sure enough, it was Halim, YB1A, in Indonesia! Knowing I probably had only one shot before the whole East Coast of the United States caught on, I carefully sent my call. A new country on 40 meters entered my log.

At this time of day, signals from Indonesia to the Southeastern United States were coming from the south, via what's called *long-path* propagation. The shorter, northerly path was completely in daylight, so 40 meter signals from that direction were absorbed by the ionospheric D-region.

My signal was pretty strong in the US, and many other American stations heard me working YB1A. Before we'd finished our contact they were tuning up on our frequency. As we signed off, a whopping pileup exploded, and YB1A showed up on the DX spotting sites. Given the limits of my antenna system, being first was probably the only way I could have worked Halim that day. I sure don't have the loudest station on 40 meters. I had to rely on my ears instead.

You easily can work 100 countries by following DX spots online or from your nearest PacketCluster node. I'm not suggesting you ignore these valuable resources. You can't monitor all bands and modes all the time. But real DXers are the ones posting the spots, not the ones following them.

How do you know whether to keep listening to a signal, or whether you should move on up or down the band? Experience helps here and doesn't take long to acquire. As you tune the band, most signals will sound about the same. For example, on the day I worked YB1A, European signals were just starting to come through and were not very strong. YB1A was much stronger, as strong as many US stations I could hear. Yet his signal was modulated by the shifting ionosphere along the 14,000 mile (23,000 km) path between us. He was stronger than the Europeans, and more fluttery than domestic signals — that's what set his signal apart and made me stop to listen.

There are two kinds of rare DX. Obviously, DXCC entities with no Amateur Radio population that are activated only occasionally by DXpeditions are rare. Except when they're on the air, of course.

There are many countries, though, that do have resident amateurs and that are still rare enough to warrant special attention. As it happens, many of them are located in Central Asia. From North America, the short path to those locations goes through the polar regions. Earth's geomagnetic field terminates at the poles, concentrating ionization that affects signal quality and strength. The same effects create the visible auroras, in this case the *Aurora Borealis*, also called the Northern Lights (the southern aurora is called the *Aurora Australis*). The ionization can be so severe that signals are completely absorbed. When signals do get through, the shifting ionization modulates them, adding an ethereal echo and flutter.

Signals coming through polar regions are often much weaker, as well as having that *auroral flutter*. You'll have to work harder to copy them, but it's worth the effort. More than once I have worked Asiatic Russian and Central Asian stations on 80 meters that everyone else ignored. Sometimes the openings lasted only a few minutes. They were long enough for one or two contacts, and then they were gone. You won't work these stations by following DX spots. By the time someone spots them they have probably disappeared. If they haven't, you'll have plenty of competition. The best way to succeed in a pileup is to get there before it starts.

Each ionosphere-Earth bounce reduces signal strength. But even nearby signals can be weak. Closer-in signals may arrive via several propagation modes. One part is reflected in the normal fashion off the ionosphere; another may travel across the ground; and a third component may be *scattered* or reflected from irregular parts of the ionosphere or from the ground itself. Sometimes it's hard to tell a nearby signal affected by scatter from a signal coming from a long distance away. With practice the subtle differences are easier to identify.

Not that the best DX is always weak, but it pays to notice how one signal sounds different from most others on the band. I recall hearing HZ1AB, a now-defunct station in Saudi Arabia, on 30 meter CW one night. They were louder than many Europeans, because they were running 1500 W into a beam antenna. But because they were significantly farther from me than Europe, their signal faded more than European signals. Ignoring my S meter for a moment I stuck around to catch their call. That was an easy one to work, but it illustrates my technique: "Does this signal sound like every other signal on the band, or is there a difference? If there is, why?"

One important point worth remembering: When you hear a pileup, you've probably found a DX station — but it's too late. The DX has probably been posted on the spotting networks and the pileup will grow and grow. You may eventually get through, but it's much easier to be the first to find the DX. The best DXers lead the pack; they don't follow it.

As Time Goes By

As daylight and darkness take turns moving around the earth, the various bands open and close. More on this in Chapter 7, Propagation. But for now, make notes about what DX call sign prefixes you hear and when you hear them. Just before sunrise and just after sunset on one end of a path, signals are often greatly enhanced for several minutes. A propagation-prediction program can tell you a great deal about these phenomena, but it is still far better to learn it through experience.

You should learn when to expect certain parts of the world to appear on your bands of choice. If your experience tells you the Europeans should be loud and they aren't, either the band is dead or your antenna fell down (it's amazing how many people find that out the hard way!). If your antenna's working, spend some quality time with your family and come back for another listen later on.

Where to Listen

Assuming you're listening on a band that offers propagation to DX locations, on what parts of that band should you concentrate? While it's true that DX stations often favor the lower-frequency parts of the bands, you should begin by scanning the entire subband for the mode you're using.

Most American HF bands are allocated according to license class. If you're serious about DXing, I strongly suggest you upgrade your license to Amateur Extra class. You might try listening just below the frequency limit on which you are entitled to operate. A DX station operating there may be listening a few kHz higher and may hear your call. If you determine that the DX station is only listening on frequencies where you aren't authorized to transmit, at least you know the band is open to that part of the world. Look for another station to work.

When you first sit down for a session at the radio, however, I suggest that you tune across the entire subband to get a general feel for what's happening on the band. You may want to have a look at other modes as well. If your main interest is CW, scan the SSB subband while you're at it. If all the signals are weak you may have better success on CW, but you'll know not to expect strong signals.

Except during contests, most DX stations prefer to operate in what are the American Amateur Extra class subbands. In some parts of the world, amateur SSB operation is permitted on frequencies not authorized for American radio amateurs. If you haven't upgraded yet, you'll still find plenty of DX to work elsewhere. Many DX stations who use SSB will call CQ on frequencies authorized for their use but not for Americans, and announce that they are listening in an American subband. Make sure you're authorized to transmit where they're listening, and make sure you have turned on the *split* function of your rig, which lets you listen and transmit on different frequencies. You don't want to receive a QSL card from the FCC!

On 160 meters, most DXers use CW, except during SSB contests. It's hard to be loud on 160, and CW's improved signal-to-noise ratio makes a difference. Radio amateurs in some countries are not allowed to operate in the segment 1800 to 1810 kHz. The usual DX window is above 1820 kHz, except in contests, when stations spread out over a larger slice of the band. Some countries do not allow Amateur Radio operation above 1825 kHz, making for a narrow *DX window*. DX stations sometimes show up below 1820 kHz, too, of course.

Later on we're going to learn about operating in contests, the best way to work a bunch of new ones in a short time. Propagation conditions vary throughout the year, but if you spend a few hours each day of the week before a contest listening around the bands, you'll have a better idea of what times will be most productive for you.

Propagation conditions also don't change dramatically from one year to the next. Keeping an almanac of what you hear at certain dates and times will be helpful next year, too. Knowing more about all aspects of radio communication not only helps you increase your DXCC total, it adds more dimensions to your experience.

More on Ear Training

I guess you can work 100 countries and have *arm-chair* copy on all contacts, but as I've said, the good stuff is often weak and hard to hear. I *always* use headphones; I don't own an external speaker. I'll say more about headphones in the next chapter, The DXers Toolkit. If you have headphones, use them.

Plan to spend some time listening for training purposes, and not necessarily looking for stations to work. Start at one end of the subband and try to identify every signal you hear. Before you hear a call sign try to guess where the station might be. Is it a domestic station, weak only because it's so close to you, or is it far away? What characteristics of the signal made you think that? Were you right? While you're working on this, you'll also be training your ears to be more sensitive to information and less sensitive to noise.

The best filter money can buy is the one *between your ears*. Experiment with the filters in your radio too, and an external audio filter if you have one, to see what combinations work best under different conditions. You may find, for example, that wider bandpass filters sometimes make signals easier to copy. Push the buttons and tweak the knobs on your radio. The more signals you listen to and the more you know about operating your equipment, the faster you'll be able to identify a needed station and put it in your log.

You probably like to work at least a few stations every time you get on the air. After you've scanned the band, if you haven't heard anything you need it won't hurt to work a few stations you don't need. Getting a few signal reports will tell you how well your station is being heard and it's always nice to make new friends. As part of your training program, though, I suggest you call weaker stations, not stronger ones. You'll have to work harder, and that will improve your skills.

What's the DX Trying to Tell Me?

Finding the DX station is only the beginning. You must get it in your log. This requires a different kind of listening. If the station is calling CQ or CQ DX, so far, so good. But if the station is saying something like "Europe only" ("EU" on CW), and you're in North America, you have a decision to make. You shouldn't call now. Should you stick around and wait for the station to start working North America, or look elsewhere?

My advice is to note the frequency for use later on. Plug it into an unused memory if your transceiver has that option, then move on. The more stations that need that DX, the more stations will congregate around it. That leaves all the other DX on the band available for the rest of us. If you pick your operating times and bands carefully, until you have about 150 countries logged, you will almost certainly find a new country every time you go on the air. While most of the world is mesmerized by the rarest one, there are plenty of other lonely DX stations who want to work you. You'll find them a few kHz up or down the band, calling CQ and getting few answers. This is the beginning DXer's dream come true! In fact, the best times to be on the air are when DXpeditions to rare countries are operating.

If the DX station is working North America, though, you may want to try to work it. In that case, your next step is to determine where the DX station is listening. Ninety-nine percent of DX contacts are made with both stations transmitting on the same frequency. When signals are weak or the DX station has a large number of callers (a *pileup*), you have to know where to transmit.

Sometimes the DX station will give a clue. "Listening 14.250 to 14.260" on phone; "Up 10," or simply "Up" on CW or digital modes. It's amazing how many stations miss that clue and call on the DX station's frequency. This usually causes an outburst of "UP! UP!" from other stations, which only makes matters worse.

Operating in pileups is a subject worthy of its own chapter and I've given it one. If you're impatient, skip ahead to Chapter 6, Working Pileups. I warn you though, patience is a great attribute for a DXer and there's no time like the present to practice it!

Chapter 3

The DXers Toolkit

A ssuming you already have a station, what other accessories and operating aids will help you chase DX? This chapter covers equipment and software I've found helpful. It isn't all-inclusive and I am reluctant to recommend one product over another, as many choices are purely a matter of personal taste.

Clocks and 24-Hour Time

Of all the gadgets in your shack, nothing is more important than an accurate, *24-hour* clock. DXers *always* keep their logs in 24-hour format, and *always* use Coordinated Universal Time, abbreviated UTC. UTC used to be called GMT, for Greenwich Mean Time, the time at the Greenwich Meridian, or zero degrees latitude. For our purposes, GMT and UTC are the same; the difference is that UTC is determined by a precise atomic clock.

No matter what time it is at your location or anywhere else on the planet, it's always the same time in UTC. I live in the Eastern Time Zone of the United States. When I worked JA2ZJW at 6:13 in the morning and the date here was Sunday, February 19, in Japan it was 8:13 *Monday night, February 20*. Fortunately, for both of us it was 1113 UTC, February 19. When you have to look up a contact in your log, you'll be glad that we're all living in the same time zone.

You can determine what time it is in UTC by referring to several Web sites, or by listening to a time-standard station such as WWV or WWVH. If your area uses Daylight Savings Time (DST), you have to readjust your thinking twice a year. As I write this my local time is 11:33 in the morning, but because I'm in a DST zone, it's 1533 UTC. When my area changes to Standard Time (ST), at 11:33 in the morning here, it will be 1633 UTC.

The easiest way to avoid confusion is to have a 24-hour clock set to UTC. Note that you still have to calculate the date. If you live in the Western Hemisphere, the UTC date changes 4 to 8 hours sooner than your local date. If you live east of the Greenwich Meridian (in the United Kingdom), the opposite occurs: Your local date may be ahead of the UTC date. For example, when it is 9:01 in the morning on November 2 in Tokyo (which is 9 hours ahead of UTC), the UTC is one minute after midnight (0001 UTC), *November 1*.

A 24-hour clock with a date function, or a computer-logging program that automatically makes UTC-to-local-time corrections, will prevent errors that otherwise might cost you a QSL card if the other station doesn't realize your mistake. I use a simple, battery-operated 24-hour clock (**Fig 3-1**). It won't go haywire if the electric power fails, and only needs adjustment every few months. My computer logging program knows my latitude and longitude, and automatically determines the UTC time and date from the *Windows* clock. I use a National Institute of Standards and Technology (NIST) application

Fig 3-1 — A simple 24-hour clock set to UTC is one tool every DXer should have.

Fig 3-2 — For contesting I use the *MS-DOS* version of *CT*, which runs well on a vintage *Windows 98* notebook computer. For radio control and keying your PC must have serial and parallel ports, missing on most modern computers.

called *NISTIME*, available for free download from many sources, to update my PC clock every few days.

Computer Logging Programs

Two types of computer logging programs are of interest to DXers. One type will handle non-contest contacts, and may support some popular Amateur Radio contests as well. The other is specifically designed for contesting only. Options and ease of use vary, as do prices. Some are free, some are rather expensive. For contesting I still use the free *CT* program, available from **www. k1ea.com/** (**Fig 3-2**). *CT* and other contest programs use prefix databases to determine where each station is located. Updated databases are available from **www.country-files.com/**.

Contest-logging programs are limited in features you may want for non-contest operating. General-operating programs provide the option to enter almost any kind of information you could thing of about each contact (**Fig 3-3**). Most of them will print labels for your QSL cards, too. This is where personal taste comes into play.

Try as many programs as you can, as most will allow you to use them for a limited time without paying. One option the program you choose must be the ability to produce log files in Amateur Data Interchange Format (ADIF). This is one option you can't do without. Log files submitted to ARRL and many other organizations, and especially log files for use with ARRL's *Logbook of The World* (LoTW) must be converted to ADIF. Log files submitted for Amateur Radio contests must be in *Cabrillo* format. Make sure your program can import and export files in both formats.

Propagation Prediction Programs

Propagation-prediction programs are very useful for DXers. Not only will they give you hints about what times the bands will be open, they are excellent educational tools. These programs fall into two general categories. Both require you to enter your latitude and longitude, a date and time, and

Rec#	Call	Date	Bnd	Mode	Pwr	On	Snt	Rec	Country	Name	Other	S	R	Comments
1367	R1ANF	19-Mar-06	40	CW	500	10:36	599	599	South Shetland Is.					Last record sent to LoTW
1366	JH4UYB	19-Mar-06	40	CW	500	10:14	599	599	Japan				R	
1365	VK6AU/P	19-Mar-06	40	CW	500	09:55	569	579	Australia					
1364	JA1NUT	19-Mar-06	40	CW	500	09:48	569	589	Japan	Shin				
1363	PR7PO	19-Mar-06	40	CW	500	09:17	599	599	Brazil					
1362	VK2BJ	19-Mar-06	40	CW	500	09:17	599	599	Australia	Barry				Sydney
1361	LU5WZ	19-Mar-06	40	CW	500	09:11	599	599	Argentina					
1360	S58A	19-Mar-06	40	CW	500	09:09	599	599	Slovenia					
1359	LU7YS	19-Mar-06	40	CW	500	08:57	599	599	Argentina					
1358	SM7GVF	19-Mar-06	40	CW	500	08:55	599	599	Sweden					
1357	OA4/DJ3...	19-Mar-06	40	CW	500	08:47	579	589	Peru					
1356	JM7OLW	19-Mar-06	40	CW	500	08:42	549	529	Japan	Suk				
1355	OL8M	19-Mar-06	40	CW	500	05:50	599	599	Czech Rep.					
1354	4U1ITU	19-Mar-06	40	CW	500	05:45	599	599	ITU HQ Geneva				R	
1353	OH0Z	19-Mar-06	40	CW	500	05:37	599	599	Aland Is.					
1352	ZL4BR	19-Mar-06	40	CW	500	05:31	599	599	New Zealand					
1351	SO5O	19-Mar-06	40	CW	500	05:26	599	599	Poland					
1350	UA2FCC	19-Mar-06	40	CW	500	05:19	599	599	Kaliningrad	Igor				
1349	OH2PM	19-Mar-06	40	CW	500	05:12	599	599	Finland					
1348	CU3W	19-Mar-06	40	CW	500	05:05	599	599	Azores				R	
1347	SV1ENG	19-Mar-06	40	CW	500	05:01	599	599	Greece					

Fig 3-3 — One of many popular general logging programs is *ACLog*. Make sure the program you choose can import and export files in Cabrillo and ADIF formats. (See text for details.)

information about solar-terrestrial conditions (see Chapter 7, Propagation). One type uses this information to perform extensive calculations. The other uses a large database of historical propagation information. The second type is a little easier to use and is good enough for most purposes.

One shortcoming of all propagation programs is their inability to accurately predict 160 meter propagation. This is not a fault on their part. Rather, it is caused by our inadequate understanding of all the factors that affect propagation on that band. As many modern transceivers include the 6 meter band, you might wish the program could make predictions for that band as well. In fact, many of them can, sometimes. These programs will tell you the *Maximum Usable Frequency* (MUF) between points of interest, based on solar activity. Although they will not produce predictions specifically for 6 meter, if the predicted MUF is close to 50 MHz it's worth turning on the radio. On the other hand, there are many, many 6 meter openings that won't be predicted by the programs, because of the variety of propagation modes available on that band. For now, the available programs are most useful on the bands between 3.5 and 30 MHz. Two popular propagation-prediction programs are discussed in Chapter 7.

The Station Computer

I use a separate PC for logging, an older IBM-compatible notebook. I have found that CRT-type PC monitors (and desktop PCs themselves) often radiate interfering signals that are hard to eliminate. Even the best contesters will not overstress the processors of older PCs. My 300-MHz laptop works very well and doesn't generate any audible spurious emissions. Best of all, it still has the now-obsolete parallel (LPT) and serial (RS-232) ports used by logging programs to key and control transceivers.

I don't mean to shortchange other operating systems, but most popular Amateur Radio software was written for the *Windows* environment. Before investing in a dedicated system, make sure it supports your software and hardware requirements.

World Atlas

DXing transforms you into a citizen of the world. I find that most DXers have a higher-than-average understanding of geography and world events. Let's step away from high technology for a moment and talk about a real book, the kind you can hold in your hand. It's possible to find all kinds of maps online now, but there's nothing like a big, heavy printed atlas. Choosing the best one requires some research. Many DXCC entities are lonely islands in the middle of vast oceans. Locating them on a world map gives me an injection of romanticism, visions of creaky wooden ships plying the waves in search of adventure. Yet some atlases ignore those very islands, or fail to locate them in relation to the rest of the planet. A good atlas also lists the latitudes and longitudes of major cities, a big help when plotting your own DX adventures. Yes, that information too is available online, but it isn't always as easy as looking it up in a book.

I still use a Rand-McNally *Great Geographical Atlas* that dates back to the last century. The world has changed enormously since it was printed, but frankly, that's part of the appeal. Since it was published some countries, like East Germany, have fallen from the *DXCC Countries List*, but many others have been added.

If your computer monitor screen is large enough, I suppose a software atlas would work just as well. The *National Geographic* sells an excellent map collection on CD-ROM. ARRL sells a *Radio Amateur Map of the World*. It's size (26 × 34.5 inches, about 750 cm ×1 m) makes it suitable for hanging on your shack wall, and it includes call sign prefixes as well as country names. You'll probably still want individual country or regional maps on which you can locate the stations you work.

DX Spotting Networks

DXers need to know when a country or entity they need comes on the

```
[US]  [3.5/3.7MHz] 1451Z = 2351JST (Last refreshed)
W6UC      1346Z   3799.0 cq dx lp              N6XT
XF4DL     1246Z   3502.0 qsx up 2              K5SWW
VK5ZSA    1232Z   3799.0 Jim vy weak this AM   AG6X
XF4DL     1212Z   3502.0 QSX 3504.5 599 VA     KT4U
VK2AIR    1153Z   3791.0 workin east coast     W6UC
N2RK      1149Z   3791.0 cq cq dx              W6UC
XF4DL     1150Z   3502.0 QRX 3504.71           KE1F
```

Fig 3-4 — DX spots obtained online from **www.dxscape.com/**. It's better to find the DX before it shows up on a spotting network to avoid the pileup that usually follows a spot.

air. Starting with the telephone, DX information networks grew in size and complexity. With the advent of Amateur Radio repeaters in the 1970s, then by means of packet radio, and now via the Internet, DX *spotting networks* have kept pace with technology.

PacketCluster networks operating on Amateur Radio VHF and UHF bands still exist, but now you can obtain the same information through numerous Web sites. I won't list specific Web site addresses because they come and go. Use your favorite search engine to look for "dx spotting network." Most sites offer several ways to display recent DX spots. DX spots include a minimum of information (see **Fig 3-4**). This includes the spotted station's call sign and frequency, the time the spot was posted and the call sign of the station who posted the spot. Now that a US station with a W6 call sign may be located on the East Coast, and a W1 may be on the West Coast, you may not be able to hear everything you see posted on the networks. (Knowing a little about propagation is a big help.)

If you live in North America you may prefer to see only spots posted by other North American stations. Within those spots you may be interested only in those for a particular band or mode. Or you can choose to view spots from the whole world. (It's always nice to see your own call spotted by a DX station!)

If you can't always be near your radio, online DX spots are good indicators of band openings. Just don't depend on them to tell you when something you need to work is on the air. If the spotted station is rare, the spot will instantly produce a pileup. Or the station may have already moved to another frequency. Very often a spotted DX station will have answered the CQ of a North American station. After the contact the DX station will keep tuning. You know the station *may* still be on that band, but you still have to find it.

Beware of typographical errors in DX spots! Some stations are in too much of a hurry to be the first to spot a DX entity. If you work the spotted station, make sure you copy the call sign correctly; don't depend on the spotted information.

Spotting networks should be part of your toolkit, but you should still spend as much time as possible tuning the bands. DX is much easier to work *before* it gets spotted; and being the first person to discover a station is like being an explorer, not just a tourist.

Chapter 4

The DX'ers Shack

Radio amateurs love gadgets and some of us never seem to have enough of them. After more than 40 years of DXing, though, I find a simple station with a minimum of equipment is better than a room full of gear. My current station consists of a transceiver, amplifier, keyer, audio filter, antenna switches and a noise-canceling device that I find helpful on the low bands, where power-line noise is sometimes a problem. Oh, yes, and a notebook computer for logging. Let's review the equipment piece by piece before we talk about how we'll arrange it in the station. Antennas have a chapter all their own, Chapter 8.

Transceiver

The transceiver is the heart of any Amateur Radio station. This is the piece of equipment with which you will spend most of your operating time, so choose one carefully. Alert readers will notice that my transceiver is more than 10 years old. What?! The author is not using the latest-and-greatest radio on the market? No, I'm not, but my radio, a vintage Kenwood TS-450S, has served me well. It was inexpensive, doesn't take up much room on the table, and is simple to operate. I wouldn't mind having something newer and more feature-laden, but I don't think a better radio would have put more countries in my log. And that's what matters. Note that in this chapter I am not endorsing any particular brands. I've used transceivers and numerous accessories made by all the major manufacturers.

Transceiver Features, Necessary and Useful

Your transceiver need not be one of the latest models, but there are some features you can't live without.

Frequency Display: Analog or Digital?

This one is easy. You want a digital display. They're much easier to use, and older radios with analog frequency displays usually lack other necessary performance features.

Dual VFOs and Memories

This is often a misnomer, as most transceivers don't have two VFOs. They generate frequencies in digital synthesizers, and what looks like two VFOs is really one synthesizer with two memories, called Frequency A and Frequency B. You can switch from one frequency to the other by pushing a button, but you can't listen on both frequencies at the same time. Some expensive radios do have two synthesizers (and two receivers) and those can be helpful features if you can afford them. But the straightforward ability to listen on one frequency and transmit on another (referred to as *working split*) is all you'll really need.

Older transceivers could not work split unless you added a remote VFO. If your transceiver works that way and you have the remote VFO, by all means use it. If you're in the market for a transceiver now, however, get one that incorporates split-frequency operation in one box.

Any transceiver allowing split-frequency operation and digital frequency display will probably include frequency memories. I find these helpful when scanning the band. If I hear a station I may want to work, but who is either in conversation with someone else or running a pileup, I put the frequency into a memory for later use.

Most transceivers, whether they are synthesized or not, have receiver-incremental tuning (RIT). This feature was included for times when a station answering your CQ is not exactly on your frequency. You can center the signal in your receiver passband without changing your transmitting frequency. If a pileup is close to a DX station's frequency it is possible to use the RIT in place of dual VFOs, but it is a poor substitute. If your transceiver offers dual-frequency operation it probably has an RIT control, and a similar function to offset transmit frequency, called XIT. I suggest that you use dual VFOs instead.

Some operators use the RIT control to change the CW pitch when receiving, or in conjunction with a passband-offset (*IF Shift*) control. In my experience, however, RIT and XIT controls don't allow fine enough tuning adjustment, and you may find you have inadvertently shifted your transmit frequency outside the range where the other station is listening.

Frequency Coverage

Newer transceivers usually operate on the 160 meter band, and many allow operation on 6 meters or even higher VHF or UHF bands. I would opt for 160 and 6 meter coverage if possible, or at least 160 meter operation.

Internal Filtering

Until recently, all transceivers were sold with a minimum of intermediate-frequency (IF) filters installed. As the process of generating a single-sideband (SSB) signal required a filter, you got one of those, but no additional filters suitable for CW or digital-mode operation. High-end modern transceivers often come with a full complement of filters, and may include digital-signal processing (DSP) filters as well. Even if you only plan to operate SSB, if your transceiver has the option of adding an additional filter, it's a worthwhile investment. Add-on filters are available from the manufacturers, but aftermarket filters, such as those made by International Radio, often provide even better performance. Some filters plug in; others must be soldered onto circuit boards inside your radio.

Filters are available in a variety of bandwidths. For SSB operation the standard bandwidth is about 2.4 kHz. That's good enough for casual operating, where you want to hear what the other operator's voice sounds like. For serious DXing, consider a narrower filter instead. They are available in bandwidths of 1.7 or 1.8 kHz, a big help in marginal conditions or in heavy interference.

For CW operation, a commonly used bandwidth is 500 Hz, with 400 Hz and 250 Hz bandwidths as options. A 250 Hz filter is a big help sometimes, but for tuning the bands I like something wider. This is a personal choice and the only way to decide is for you to try both. If your transceiver allows you to bypass a second filter, you can use a 400 or 500 Hz filter in the first position and a 250 Hz filter in the second.

Internal DSP filters vary in performance. If your transceiver can use optional IF filters as well, by all means install them. I find DSP filters most useful for SSB operation.

Separate Receive Antenna Jack

If I wrote the rules, every transceiver would come equipped with a separate receive-antenna jack. There are other important features you should consider when buying a transceiver, but this is a good one to have if possible. If you operate on the 80 and 160 meter bands you will likely want to use a separate receiving antenna. Directional transmitting antennas for those bands are huge, but directional receiving antennas are not. If, like mine, your transceiver does not have a separate jack for a receiving antenna, you can build a switchbox to do the job. I used surplus relays designed for radio applications, found on an online auction site. The amplifier-control relay in my transceiver switches this box when needed, along with my amplifier and noise-canceling device (see below).

Internal Keyer

Some transceivers have internal CW keyers, or even memory keyers. I wouldn't base my decision on that feature, as external keyers and computer programs are easy to connect and use.

Overall Performance

I've saved the best for last. The most-important part of your station is the receiver. While any receiver will hear signals during casual operation, a DX pileup or contest will drive lesser radios right off the band. The selectivity of a receiver is ultimately determined by its internal filters, but they may very well be located too far downstream to prevent off-frequency signals from getting in and messing up reception. For example, if you were listening on 7200 kHz with a pair of good 1.8-kHz filters, you would not expect to hear a signal at 7190 kHz. You won't, but if that signal is strong enough it can overload earlier stages of your receiver, reducing the sensitivity and generating noise that will wipe out the signal you're trying to hear.

QST Product Reviews go to great lengths to quantify this information based on standardized tests conducted under laboratory conditions. This is a place where anecdotal user information is often unhelpful. Too often I hear users proclaim the *sensitivity* of their receivers, where sensitivity is usually the least-important factor.

Sensitivity is properly defined as the ability to hear weak signals. Under laboratory conditions, where there is no atmospheric or man-made noise to contend with, good sensitivity looks like something we'd all want the most of. Except perhaps on the 12 and 10 meter bands, external noise will wipe out any advantages of extremely high sensitivity. That's because sensitivity is largely affected by noise generated within the receiver. If the external noise level exceeds that generated inside the receiver, it becomes the determining factor in whether you can hear a station or not. Decreasing your receiver bandwidth reduces all noise, internal and external, which is why we install optional filters whenever possible.

The parameter of most interest in the DXer's receiver, however, is its *dynamic range*. One measure is the receiver's *blocking dynamic range*. ARRL injects two signals into a receiver, a weak one at the operating frequency, and a strong one offset by several kHz. The operating-frequency signal level is left constant while the off-frequency signal level is increased until the former level appears to decrease. That is the point at which the receiver begins to *block*. On the air you would hear this as a *pumping* of the desired signal. Sometimes the desired signal will just disappear.

Another measure of dynamic range is related to *intermodulation distortion*, often abbreviated IMD, in a receiver. This is a rather more complex phenomenon than simple blocking. Two equal-strength off-channel signals spaced between 2 to 5 kHz apart are applied to a receiver, and their level is adjusted until non-linearity in the receiver's front end causes a spurious signal to be produced. For example, intermodulation distortion will occur if strong signals at 14,020.0 and 14,018.0 kHz (2 kHz apart) are applied to a receiver tuned to 14,016.0 kHz, 2 kHz away from the closest interfering signal. The level of the interfering signals is increased until the IMD is just detectable. The IMD produced within the receiver due to strong off-channel signals will produce "beeps and squawks" on the frequency the receiver is tuned to, even if there is no actual signal at that frequency.

ARRL members can learn more about how equipment tests are performed and what the tests mean, by reading articles listed on the Technical Information Service "What Rig Should I Buy?" page, **www.arrl.org/tis/info/ rigbuy.html**. Product Reviews going back to 1980 are also available to ARRL members. ARRL has modified and refined its measurement practices over the years but the older reviews will still tell you much of what you need to know. If there is a single most-important parameter in transceiver design, it is receiver performance, so pay careful attention to ARRL measurements when deciding what to buy.

I haven't mentioned transmitter performance, and of course that's important too. In general, transceivers with good receivers tend to have good transmitters. Your transmitter should have voice-operated transmit (VOX), which is useful on CW as well as SSB. Built-in speech processing is very useful, but don't overdo it. The quality of transmit audio is important, and something too few radio amateurs take seriously enough.

If CW operation interests you, the transition time between transmit and receive is very important. When you hit the key the transceiver has to switch into transmit mode. The internal frequency synthesizer may have to shift frequency if you're operating split, and the antenna must be transferred to the transmitter. If you're using an external power amplifier, the relay that controls the amplifier also has to move. When you stop transmitting, everything has to switch back again. In heavy pileups a transceiver that quickly transitions back and forth is a great advantage. *Full break-in*, or full-QSK, means you can hear the other station between dits and dahs you send. While ideal for CW operation, few transceivers offer that feature. Reducing the VOX Delay helps for those that don't, but an external amplifier relay may not be able to keep up. Even if your transceiver operates in full-QSK mode, your amplifier may not. More on this later when we discuss amplifiers.

QRP Transceivers

Very few QRP (5 W output and less) transceivers provide the features I think are important for DXing. Most don't have dual VFOs or digital displays, and they usually lack the receiver performance needed in a competitive environment. Lightweight radios are great for casual operating. Serious QRP DXers would do better by reducing the power output of a modern 100 or 200 W transceiver, which will give them all the advantages available to QRO (more than 5 W output) DXers.

Voice and CW Keyers

Through the miracle of modern digital electronics, you can now record and transmit entire QSOs in any mode with the push of a button. Digital-mode operators do this all the time; only on voice and CW modes are they still a bit controversial. Here's how I look at them. Finding the DX station requires skill, knowledge and perseverance. Getting on the right transmit frequency and timing your calls takes even more. Compared with these requirements, the ability to say or key your call sign seems trivial.

Amateur Radio contesters have been using voice and CW keyers for decades, and there's no reason DXers shouldn't use them as well. Of course, we'd all like to get through a pileup on the first call, but when you can't, having a *robot helper* lets you concentrate on the essentials. I don't operate SSB very often, but if I did I'd want a voice keyer. There are two types, those that record your actual voice in digital memory, and those that digitally generate a voice that sounds like a robot. When you're chatting with a DX station you may prefer to transmit your actual voice, but the digitized voices sometimes work better in cutting through interference. If you can't soundproof your shack, a voice keyer lets your family live in peace while you're on the air, too.

If you're going to buy a CW keyer, a memory keyer is the only way to go. Several models are available. I prefer simple and small, and the four memories in my Logikey keyer are just what I need. I have my call sign in one, a simple "5NN TU" in the second, my name and state abbreviation in the third, and a short "CQ" message in the fourth. While I'm hunting for the right transmit frequency with one hand, I only have to tap a button with the other when I want to send my call or, even better, the signal report when I finally get through the pileup. In my younger days I did everything manually and managed to get by. If that suits you, by all means use your keyer paddles. When I stay up all night to chase DX on the low bands, I find the memory keyer just as important as a cup of strong coffee.

For CW contesting, use a program that sends your exchange and other information via keyboard buttons. Most programs will let you switch to *keyboard* mode if you want to chat with a friend during the contest.

Code-Recognition Accessories

If your interest turns to CW but your code speed is slow, you may be tempted by the code-recognition boxes and software on the market. I recommend you spend the money on software to help train you increase your speed instead. Some software will even simulate pileup and contest conditions, with added noise effects for good measure. While there is nothing inherently different from using code-recognition software and operating on the digital modes, band conditions are rarely good enough to allow code-recognition devices to perform as advertised. When the DX station calls you, you have to be able to copy that call yourself. Being able to copy Morse code is traditional among CW operators and is a skill well worth developing.

External Audio Filters

There is no substitute for good filtering inside the receiver, but sometimes we can use a little extra help. Over the years I've tried almost every external audio filter there is. My current transceiver has two internal CW filters, and for that mode I rarely need an audio filter. I still keep two audio filters available for occasional use. One is an all-around analog filter no longer manufactured. It has several options still offered on some currently manufactured filters. I can attenuate audio frequencies either above or below the range I'm listening to, or limit the response to a range of audio frequencies with greater or lesser selectivity. An additional notch function comes in handy when someone decides to tune up near the frequency on which I'm listening.

My other audio filter uses switched-capacitor circuits. Low-frequency attenuation is fixed internally; a front-panel knob lets me adjust high-frequency attenuation. I use both filters often enough that I'm glad to have them. If your choice is between adding internal filters to your transceiver or buying an external audio filter, get the best internal filters you can afford. External filters are nice to have; internal filters are required.

Microphones, Headphones, Headsets

Microphones and headphones play a greater role in DXing than many people realize. Let's start with the microphone. The best microphones for DXing and contesting will not relay all the nuances of your voice. By design, their frequency response is limited. Assuming your transmitter is limited to a specific peak-envelope-power (PEP) output, narrowing the frequency response increases the effective PEP. While you give up some fidelity, you gain in signal strength at the other end. The microphones supplied with transceivers, or optionally offered by transceiver manufacturers, sound nice on the air but aren't good enough for competitive use.

I don't recommend desk and hand-held microphones, regardless of their audio performance. A desk microphone makes you lean toward it, when you might wish to turn your head to look at the meters or turn your antenna. A hand-held microphone unnecessarily ties up one hand. A competition-grade headset with an integrated boom microphone is the ultimate way to go. Such a headset also incorporates headphones with a limited frequency response.

Many operators use a pair of stereo headphones. The better ones are comfortable and sound great, but that can be a disadvantage. You don't necessarily want high-fidelity reproduction at high audio frequencies. You do want comfort and low distortion, of course. But for voice operation frequency response above about 2800 Hz can be a drawback. This is because many transceivers produce a lot of high-frequency hiss. This can be tiresome to listen to over a long period of operation.

For CW operation, response much above 800 to 1000 Hz can be useless and annoying, often due to the same hiss problem. An external audio filter can tailor headphone response. Even if you don't operate on voice modes you may wish to consider a headset with an integrated microphone, since the headphone's frequency response is tailored for communication use. Check the ads in the back pages of *QST* or *CQ* magazines for headphone/microphone headsets, or visit the **www.heilsound.com/amateur/** web site of Heil Sound.

Optional external speakers are popular, but not recommended. Some do have internal filters, but unless you spend a fortune modifying the acoustics of your shack, they will never give you the clarity you get from wearing headphones. If you want to monitor a frequency while you do other things in your shack, your transceiver's internal speaker is good enough.

Amplifier

Do you need an amplifier to work DX, or is your 100 W transceiver good enough? No, you don't need an amplifier, but they sure help sometimes.

Amplifier Pros and Cons

I know many people who have worked lots of DX while running 5 W. They laugh at anyone running 100 W, much less an amplifier. Even on the 80 and 40 meter bands, I have been able to work stations on the opposite side of the world while running 5 W. Yet at other times, even 100 W would not get me through a big pileup, and I needed my amplifier. Operating skill is more important than power, but there are times when you need a little help!

The amplifier I'm using was a gift. It produces about a kW on CW and SSB. My previous amplifiers, an old Heathkit SB-230 and an even older Heathkit SB-220, still serve as good examples of how it goes when you run higher power. When I first got on 40 meters from this location, one of my first goals was to work Japan on CW. While ragchewing with CN8GL in Algeria, waiting for the band to open to Japan, the amplifier blew up. Not only could CN8GL still copy my signal, but I got a big surprise later that night. Despite being without an amplifier, I called CQ DX around the time of sunset in Japan, and the first station to call me was in Japan. So it turned out I didn't need the amplifier to work Japan on 40 meters. On other nights, though, it has been impossible to work the Far East on that band, or worse yet on 80 meters, without the amplifier.

My antenna options are limited, as I live in a condominium. Better antennas would certainly improve my chances and would also make it easier to hear other stations. That's where the tradeoff comes in. Assuming you have optimized your antenna system, having an amplifier on hand will often make the difference. Another example, the DXCC entity of Peter I Island is in the Antarctic region. I have worked a few of the semi-permanent bases in Antarctica while running 5 W, on both 80 and 40 meters. When the Peter I operation began, however, I spent several minutes trying to work them with 100 W, just to see if I could. No luck. As soon as I turned on the old SB-230, which puts out about 400 W, I got through on my first call. Had there been fewer stations in the pileup I'm sure I could have worked them with 100 W. If that's all I had, I would have kept trying and eventually gotten through.

Back in the days when I ran a Heathkit SB-220 at 1 kW I had a pair of loop antennas with a common feed point, switched with a large relay. It worked very well until the day I noticed my SWR was jumping like crazy. Inspection revealed the charred remains of the relay. High power will quickly seek out and destroy the weak links in your antenna system. At the 100 W level, if your feed line isn't too long, you can get by with *quality* RG-58 coaxial cable. When you move up to more than 400 W, you'll have to invest in bigger, more expensive cable, like RG-213.

And there is increased potential for interference to consumer-electronics devices with higher power. You may well decide you can live without an amplifier for now. But here are some tips on selecting an amplifier if you do decide to go that route.

Amplifier Selection

There are plenty of amplifiers available and it will pay to do some research before buying one. *QST* Product Reviews and user comments at **www.eham.net/** are good places to start. Bear in mind that reviews at the latter site are extremely subjective and not everyone's operating habits punish an amplifier like really heavy duty DXing and contesting. If an older amplifier tempts you, make sure your transceiver's keying circuit can switch the amplifier's relay. Unmodified Heathkit amplifier relays ran on 120 V dc, and will quickly destroy the amplifier-control relay in a modern transceiver. Aftermarket modification kits are available. Newer amplifiers can be safely keyed by modern transceivers.

The relay plays another role if you want fast transmit-receive switching. Most amplifier relays are big and hence slow, and will not follow even slow

Morse code sending. For voice VOX-type operation they are usually fast enough. My homemade amplifier will incorporate fast vacuum relays. The Ameritron AL-80B I'm using now is slow enough that if I planned to use it forever I'd probably install faster relays.

Frequency coverage is another important consideration. The 160 meter band is becoming more popular, and your amplifier should cover it. Newer amplifiers do, but older ones may not. No high-power amplifiers capable of operating on 160 through 10 meters (or 80 through 10 meters) also cover 6 meters. If your transceiver includes 6 meter coverage you'll need a separate amplifier for that band.

Most amplifiers can run on either 120 or 240 V ac lines. I strongly recommend running them on 240 V ac if you can. Some manufacturers claim their amplifiers will run as well on 120 V, but some simple mathematics disproves that. A 1000 W output amplifier requires about 2000 W of ac input power on signal peaks. That works out to almost 17 A on a 120 V ac circuit. If your shack has a dedicated 20 A, 120 V ac circuit you may get by, but if the distance from the outlet to the circuit breakers is long, you'll notice the panel lamps flickering as you transmit. On a 240 V circuit the current is less than 9 A, significantly reducing voltage drop (and heat) in the wiring.

Other Gadgets

Clock

I've already mentioned a 24-hour clock, but it's worth mentioning again. Mine is battery powered and keeps good time. I don't have to worry about resetting it after a power outage. I keep it set to UTC. There are fancier clocks but none easier to read at a glance.

Noise-Canceller

Powerline and other manmade noise is a frequent problem where I live, and is especially noticeable on the lower-frequency bands I like to use. Like audio filters, such devices are no substitute for fixing problems farther up the chain. Relocating antennas and consulting with the power company are the best ways to solve noise problems. Where I live, though, I sometimes receive spurious signals from TV sets and other devices and an MFJ-1026 *Noise Canceling Signal Enhancer* has sometimes made it possible to hear signals I would otherwise have missed.

This unit is easy to use, but you'll need a separate noise-pickup antenna. The internal antenna usually isn't large enough. In use, the signal from your receive antenna goes in one port and the noise antenna is connected to another. Your transceiver connects to a third port. You adjust the controls so that a signal picked up by the noise antenna is equal in strength but opposite in phase to the same signal picked up by your receiving antenna. For a single noise source the device works very well. If you have to contend with multiple noise sources the device may be less useful. If you're fortunate enough to have no local noise sources, you won't need this accessory.

Station Layout

Simply stated, you should be able to operate your station with a minimum of head and hand movement. You shouldn't have to stretch or stand up to reach any control you need during normal operation. The more functions you have to control the less likely you are to be paying close attention to what's coming in through your headphones. Even if you can reach every control without stretching, you should try to minimize the number of controls you have to touch. At all times your attention should be focused on listening.

My shack is minimally equipped. That's because I set it up on the porch when I want to operate and then bring the equipment inside when I'm going to be off the air for a few days. I know the best settings for my audio filters so I rarely have to touch them while operating. On 80 and 160 meters I have to switch receiving antennas and may have to adjust the noise-canceling device. But most of the time my attention is all on picking out and identifying signals — the weaker the better! Despite having a very simple and inexpensive station, I can usually work all continents on 80 or 40 meters on any given night, and have worked 100 DXCC entities in one night on 40 meters, during a DX contest. The next night I almost duplicated the feat on 80 meters.

In many hours of operating I have never wished for a more-complicated station. DXing is all about operating, not equipment. Your comfort and convenience are the most-important parts of your station.

Chapter 5

SSB, CW, RTTY, or ... ?

You may be wondering if a particular mode will make it easier to get DXCC. The answer to "What mode should I use?" is simple: Use as many as possible. The Mixed DXCC award permits operation on any authorized Amateur Radio mode. Leave the single-mode awards for later on.

Having said that, let's talk about the advantages and disadvantages of the different modes.

CW, the Small Station's Friend

With so many technological advances, Morse code (CW) is no longer the ultimate mode for weak-signal communications. For DXing, however, it still has a lot to offer.

There are still many hams on the air who like CW, and being able to work them will improve your chances of adding to your DXCC total. While some digital modes are almost as effective at low power levels or during poor propagation conditions, there are more CW signals on the bands.

CW is much easier to copy in poor conditions than a voice signal. If you have one or more CW filters in your transceiver, or even if you only have an external audio filter, you can decipher a 100 W CW signal at times when a 1000 W SSB signal would be nearly unintelligible. Your moderate-power CW signal can get through when an SSB signal at the same power level will not.

CW is an acquired taste. When 13 and 20 word per minute (WPM) speeds were required to obtain a US license, most CW fans learned to appreciate the mode by force, not by choice. If you are willing to give it a try, however, there are many computer programs available to help you. Some will even simulate pileups or contests. They often provide built-in noise sources to simulate actual on-the-air conditions. These programs are a big help when you can't get to a radio, but there is no substitute for actual on-the-air practice.

I don't recommend devices or software that decode and then display Morse code. I don't believe they help improve your code speed, and they are incapable of accuracy under marginal conditions. Depending on a code reader in a pileup will cost you many new ones.

SSB

At first glance, SSB operation seems the easiest of all. You just find the DX, push a button and send your call. So far, so good. Bear in mind, though, that as many as 20 CW stations could fit into the bandwidth occupied by one SSB signal. Good ears and good equipment allow the DX operator to pick out CW signals much more easily than can be done in an SSB pileup. On the other hand, an SSB pileup sounds like 10 to 20 kHz of pure noise, occasionally interrupted by one or two audible signals. SSB DXing requires at least as much operating skill as CW. Unless your signal is twice as strong as others calling on or near your transmitting frequency, you won't be heard unless your timing is perfect.

Digital Modes

Radioteletype (RTTY) and other digital modes were the biggest beneficiaries of the personal computer. Many stations still use RTTY, the oldest keyboard mode, but many radio amateurs are discovering other modes like PSK-31. PSK-31 is well suited for use in marginal conditions, and the software makes it easy to find and contact other stations. RTTY is just as easy to use, but doesn't lend itself quite as well to marginal conditions.

RTTY uses *frequency-shift keying*, where the Mark and Space tones are offset by a fixed amount, usually 170 Hz. That shift can cause problems when propagation conditions are unstable, as the Mark frequency may fade while the Space frequency does not, or vice versa. Modern software can handle these variations, but doesn't cope well when interfering signals appear on or near the same frequency.

RTTY receiving software simulates a frequency-modulation (FM) receiver. Incoming audio tones from the receiver are amplified and *limited* to reduce fading effects. The result, though, is that only the strongest signal in the receiver passband is received. Where a good operator can pick out two CW or voice signals on almost the same frequency, RTTY software either decodes only the strongest signal, or if the signals are close in signal strength, may not be able to decode either of them.

Something else to consider about the digital modes is the on-air duty cycle. When transmitting, the digital modes produce a constant carrier, shifted in phase or frequency, but not in amplitude. A 100 W transceiver of recent manufacture is not designed for full-time operation (a 100% duty cycle) at that power level. Likewise, most modern power amplifiers are not rated for full-time operation at their rated power output. Without an amplifier, you will be limited to about 40 W output on the digital modes. Unless rated for continuous operation at full power, an amplifier rated for 1500 W PEP output should be limited to about 600 W output.

With those limits in mind, if your shack includes a computer capable of running digital-mode software, it's worth learning to use it. Many DX stations and DXpeditions now include RTTY and other digital modes in their operating schedule.

Most software uses your computer's sound card to process received audio and to generate transmitted audio. You place your transceiver in SSB mode and connect the sound-card output to your microphone connector. The sound card can easily overdrive the microphone input, severely distorting your transmitted signal. In addition to being illegal, the distortion may make your signal unreadable. An oscilloscope designed to monitor your transmitted waveform will help you set the tone levels, and is also handy when operating SSB.

There is a variety of software available for RTTY operation. Search for "RTTY software" on the Internet to locate such programs, many of which are free of charge.

Chapter 6

Working
Pileups

While you can easily earn DXCC without ever jumping into a pileup, I'll bet that no DXer can resist one. Pileups are the economics of Supply and Demand applied to Amateur Radio operating. One station that many DXers would like to put in their logs is besieged, and tries to work as many of them as possible. Pileups are exciting, exhausting and, depending on propagation conditions and the skill of the DX operator, often frustrating. The most-important rule is that you must *listen carefully*. How many times you transmit your call in a pileup is not important. Transmitting it once, on the right frequency and at the right time, is all it takes. Quality, not quantity, is what counts. Knowing where and when to send your call sign can only be learned from careful listening.

What Is A Pileup?

If I call CQ from my US location in Florida, on a band that is open to Europe, I may get two or more simultaneous replies. I can choose one or the other, and the one I don't choose will probably move along to find someone else to work. If my call sign identifies my location as *rare*, I may be called by a handful of stations, and the ones I don't answer right away will stick around and keep trying. Once the first station gets through, he or she may post my call sign and frequency on a spotting network, alerting thousands of others to my whereabouts. About the time I work the third station the floodgates will open. In minutes the mob will grow from three stations, to a dozen, to possibly hundreds. Now I've got a pileup. So here's a hot tip: Get there early and work the DX *before* the pileup gets big.

I Was Too Late — Now What?

Even if you arrived late to the pileup, all is not lost. Before diving in, though, take time to study the situation. The worst case is when you hear a pileup spread over several kHz, but the identity and transmitting frequency of the DX station is a mystery. What can we learn from the pileup? First, note what stations are calling the DX station. If you are in North America and every station in the pileup is in Europe, you may not be able to hear the DX station. It's possible that Europe has propagation but you do not. It's also possible that the DX station is working only European stations at that moment. To find out, let's see if we can find where the DX station is transmitting.

Sometimes the pileup is concentrated very near the DX station's transmitting frequency. This usually happens in contests, where a band may be very crowded with many signals, or where the DX station cannot work split (see Chapter 4, The DXers Shack). As a rule, these kinds of pileups tend to string out slightly above the DX station's transmitting frequency, perhaps with a few stations calling below it. Tune to the high-frequency edge of the pileup and listen when the pileup stops calling. If you don't find the DX station there, keep tuning lower in frequency. Chances are you'll find the DX about 5 to 10 kHz below the edge of the pileup, sometimes less.

If you have a rotatable directional antenna, you may not have it pointed in the right direction. Use your propagation software and your knowledge of propagation to determine when propagation favors working the DX station you hear in the pileup. As this book is written during the low part of the solar cycle, propagation conditions do not favor paths through nighttime regions on 20 meters. If it is near darkness in Europe and you hear Europeans in a pileup, they are working someone either slightly north, or south or west of them (but unlikely to the east of Europe). If your antenna is pointed at Europe you may be looking in the wrong direction to hear what they are hearing. Having an omnidirectional backup antenna, like a multiband vertical, may help here, assuming the DX station is strong enough.

Note that I have not recommended checking the spotting networks. They may give you the answer, but you'll never learn anything that way!

What To Do When You've Found the DX Station

Once you can hear the DX station how do you add it to your log? What information is the DX station providing? Are you sure you've copied the call correctly? Is he or she giving instructions about where to call, or working only a certain continent or perhaps a certain US call area?

An example would be when the DX station says, "Up 5 to 10, Europe only." Now you know where to call, but if you aren't in Europe, you'll have to wait. It may seem unfair to be left out when propagation favors a contact, but it is the DX station's party. Your invitation will come, and you just have to wait for it. Look for something else to work, catch up on your QSLing, or take the dog for a walk. But don't try to crash the party.

Instructions on CW or RTTY are similar to those used on voice, but are abbreviated. Sometimes the DX station will only say, "Up," or "Up 5," or only "U5." "Up 5" usually doesn't mean to call only at precisely 5 kHz higher in frequency; it only means the DX station will not be listening for calls lower than that in frequency.

By now, the reasons for working split should be obvious. Keeping the DX station's frequency clear allows everyone to hear him without interference. Spreading the pileup over several kHz makes it easier for the DX station to pick out call signs, speeding up the process. Non-DXers sometimes hate pileups, since they consume several kHz. DX stations should never allow pileups to get too wide, to avoid overrunning other stations. They do this by not responding to calls from stations operating above or below a certain limit.

Your next task is to determine those limits. If the DX station says "Listening up 5 to 15 kHz," or, even better, "Listening 14 point two-three to 14 point two-four," that's where you have to call. Don't be distracted by other stations trying to hang off either edge of the pileup.

Spend a few minutes listening to the pileup, and also learning what stations the DX is calling. If the successful stations are in the Northeast US and you're in the Midwest, propagation may not yet favor you. Propagation conditions tend to move east to west, and your turn may come soon, so it's worth waiting. If the successful stations are all in Europe, you may have a long wait, even if you can hear the DX station. The Europeans may be much louder, and you won't be heard. If the DX station is not specifically ruling out calls from the US, decide how much time you want to invest. It may be better to look for something easier to work, and trying to work this station at a more-favorable time.

Listening in Pileups

Sometimes the pileup is spread over only a few kHz close to the DX station's transmitting frequency. These are the hardest ones to figure out. You don't want to call on the same frequency as several other stations, but you don't want to be too far away from where the DX is listening either. It's hard to resist the temptation to transmit every chance you get, but try. Instead, see if you can figure out how the DX station is tuning between contacts. Is he tuning above his frequency only, or sometimes above and sometimes below?

Timing is sometimes more important than being on exactly the right frequency. There will often be a flurry of stations calling again and again. Then, as if on cue, they will all stop to see if the DX comes back to anyone. If you can drop your call in once, exactly at the moment the rest of the pack stops to catch its breath, you may be the lucky contestant. Techniques like this require enormous self-control, which comes more easily when they've worked for you a few times already.

When the pileup covers a more-or-less known part of the band, usually above the DX station's transmitting frequency, at least you have a fair idea of where to transmit. Again, though, listen before transmitting. Experienced DX stations will often tune in predictable increments after each contact. By flipping back and forth to see who the DX just called and where that station is transmitting, you can figure out where to transmit when the time comes. You may not always be able to catch the station being worked; this gets easier with practice. If you can prove there's a pattern at work, stick to it and you'll probably be surprised at how quickly you get through.

If the DX station is randomly tuning through the pileup, probably looking for the strongest signals, there is no pattern to follow. Sometimes it pays to pick a frequency and stick to it. Other times it's better to quickly tune through the pileup looking for a relatively quiet frequency on which to transmit. You have to do this fast, though, and it quickly becomes very stressful.

Jumping Into The Pileup

Far too many stations do not understand how to *break* a pileup. They waste their time and create needless interference. I look at it this way: If I'm going to spend my time on the air I want to maximize my chances of success. Sitting on one frequency and transmitting your call sign at irregular intervals *may* result in a contact. Taking the time to study the pileup and the DX station's operating habits will greatly increase your chances. Sometimes you do get lucky; the best DXers make their own luck, however.

I've found that many, if not most, stations calling in a pileup sit on one frequency and hope to get lucky. In fact, many of them send their calls every time the DX station is listening, even if the DX station just called someone else. This is an unfortunate byproduct of a time-honored technique called *tail ending*. Tail ending worked better in the days when DX contacts were longer than a simple exchange of signal reports. The non-DX station would conclude with "73, this is ...." and a sharp DXer would drop his or her call sign *once* at that exact moment. This can still work if your signal is strong enough to ride over the top of the other station's signal, but it has become common practice and is too often abused. Now some people send their calls repetitively while the non-DX station is trying to send his or her exchange (the DX's signal report and perhaps the non-DX station's location).

This is very bad operating practice and serves only to reduce the DX station's contact rate. Having to constantly ask for a repeat due to unnecessary interference does no one any good. The DX station may get fed up and leave the air.

So don't transmit just because you can. Transmit only when you know the DX station is listening for new calls, and try to choose a transmitting frequency where you expect the DX station to be listening. Otherwise you're wasting your time and energy. DXing is a sport, not a brawl. We may be amateurs but that doesn't mean we can't act like professionals.

Pileups centered on the DX station's frequency are the hardest to break. Getting through often requires a combination of brute force and sheer stubbornness, mixed with a good sense of timing. With many stations calling close to the DX station's frequency, it's hard to know when he or she calls someone. So the pileup blazes on, with stations repeating their calls over and over, until they seem to take a collective breath and everyone stops to listen. After a second or two, if the DX does not call anyone, the madness resumes.

Sometimes being the last station to call — after everyone else has paused — will do the trick. Unfortunately, as soon as one station breaks the silence, everyone else jumps in again. Timing your calls, and trying to find the least-busy frequency within the small area in which the DX station is listening are all you can do.

I have to take down my stealth antennas during the daytime, and they need to be out of sight before my neighbors are out and about. During a DX contest last year, I decided that I would start taking them down at 6 o'clock in the morning, local time, no matter what. When the clock rolled over to 1100 UTC, I would shut down and head outside. As luck would have it, an Australian station showed up on 80 meters not long before my deadline, and I needed it for a contest multiplier. His signal was not very strong, and everyone else seemed to need him too. The pileup was brutal and chaotic. My clock was already displaying 1059 when there was a brief pause in the pileup and I sent my call once, very fast. As the clock turned to 1100 I was logging the contact. In this case I would have to credit perseverance and confidence, more than technique. I knew he could hear me, if only I could get my call sign in when the pileup was catching its breath. I wasn't the strongest signal in the pileup by any means. I hope everyone else worked him, but I didn't stick around to see. A minute later I was untying ropes and coiling up wires... and smiling.

Pileups spread over several kHz, where the DX station's frequency is clear, are much easier on the nerves. They still pose a few problems for the DXer, who has to decide where to receive within the pileup. DX stations may tune randomly through the pileup, picking calls more or less at random. Those pileups are harder to break. One school of thought says you may as well pick one frequency and stick to it. If you are strong enough, eventually the DX station will hear you.

I say it depends on how dense the pileup has become. Even when many stations are calling, you will often find a few unoccupied kHz. Of course, what sounds unoccupied on your end may not be so quiet at the DX station's end, but if you can hear most of the calling stations, you may be able to find a clear frequency. If the pileup has reached critical mass, pick a frequency and hope. Send your call quickly, one call at a time. Don't call over and over. Call once and listen. Then try again.

Should you hang out at the high end, low end or in the middle? That's a tough one. I usually keep moving, looking for quiet spots. I've never noticed that a big pileup is quieter at either end or in the middle. Parking on one frequency and sending my call over and over is beyond my patience and I avoid it.

My transceiver has a button that allows me to momentarily listen on my transmitting frequency. When I release the button I'm hearing the DX station's frequency again. I have done well by pushing the button with my right index finger and tuning quickly through the pileup with my left hand. If I find a clear frequency I release the button. If the DX station is not transmitting I send my call *once*.

There are two problems with this method. One is, the DX station may already have answered someone else. They tend to set a rhythm, though, which gives you a clue as to how much time you have to seek out a clear frequency and send your call sign. It may be only a second, and that can lead to the second problem. If your tuning knob is still moving when you release the button, your transceiver can zip right off the DX station's frequency! Make sure the knob has stopped turning before releasing that button.

Saner conditions prevail when the DX station tunes in predictable increments through the pileup. In either case, you'll have to find and identify several of the stations the DX station is working before you know. If the DX station is moving up in, say, 1 kHz increments, I try to call just a bit above the next increment on CW, or close to it on SSB. You can't be too far off on either mode or you'll be outside the passband of his or her receiver.

There's another part to this technique. When DX stations reach the upper edge of a pileup, they may start tuning down in the same increments, or return to the bottom edge and start tuning up in frequency again. This method keeps you busy, but I always get through faster when I have a clear idea of where the DX station will be listening for the next call.

It's hard to stop and listen when you really want to send your call sign. You'll never work the DX if you don't send your call sign, right? Bear in mind that you will never work the DX if you send your call sign at the wrong time or on the wrong frequency either. Patiently studying the pileup and developing a strategy is not a waste of time. Rather, it will save you time and improve your chances of making the contact.

It at once aggravates and amuses me to hear many stations broadcasting on the same frequency over and over, when I have slipped into the same pileup and made the contact right under their noses. These would-be beacons create a lot of interference, but I suppose they help keep the electric power utilities in business.

The *Partial-Call* Syndrome

DX nets and lists have produced some bad voice-mode operating habits. One of them is sending only part of your call sign. Don't do it. Send your full call every time. The DX station may only catch a piece of it. If you send "One

Sierra" instead of your full call sign, though, you end up wasting everyone's time, because chances are the DX station will be able to copy your full call sign.

Using Phonetics

When using a voice mode, we all know we have to send the letters phonetically. There is a standardized phonetic alphabet, but don't get locked into it. People whose primary language is different from ours may not understand every phonetic, especially when conditions are poor. If a DX station calls you but has only gotten part of your call, try changing phonetics. I heard a station on 75 meters recently, being called by a DX station who was missing the last letter of his call sign. He kept repeating the same phonetics over and over again. Good thing I didn't have my VOX turned on, because I was yelling "Try a different phonetic!"

If "Kilo Romeo One Sierra" doesn't work, I'm ready to try "Kilowatt Radio One Santiago" or other variations as needed. It helps to know your geography. If one letter of your call is miscopied or not copied at all, and there is a major city in your target's country that starts with that letter, give it a try. If you're trying to work a Russian and "Mike" doesn't work, try "Moscow" (pronounced 'Mos-ko,' not 'Mos-cow'). Be creative but don't get carried away. "Sierra," "Santiago," "Stockholm" and even "Sideband" are okay, but "Sidereal" or "Synchronicity" are pushing it, unless you know the DX station is a New Age astronomer.

Foreign Languages

Knowing a few phrases in some foreign languages can be a big help. Most DX stations know a little English, but they'll appreciate your efforts to at least say "Good-bye" in their language, if you don't mangle the pronunciation too badly. I have an old phrase book for radio amateurs, which I believe is out of print. I'm surprised there isn't enough interest among the DXing community to support a book like that. Speaking to other stations in their own language helps return some humanity to what often seems like all-out electromagnetic warfare.

Advice For DX Stations

The DX station is the master of ceremonies and the commander in chief of the pileup. When pileups degenerate into anarchy, it isn't necessarily the fault of the DX station, but there are ways they can maintain better control. That not only makes DX chasers happier, it makes life easier for the DX station, too.

The burning question around the DX station's frequency is always, "What's his call?" Next up is, "Where's he listening?" If the DX station is trying to keep a clear transmitting frequency, the last thing he or she needs is a bunch of half-crazed DXers sharing information right on top of it. I listened to a major DXpedition for at least an hour last year, during which time I never heard the operator send their call sign.

Yes, it was on every spotting network in the world, but they should be adjuncts to good operating, not substitutes for it. It only takes a fraction of a second to send your call sign after each contact. Worst-case, send it after every five, or at most, 10 contacts. At the same time, specify where you are listening, and stick to those limits. If you say you are listening up 5 to 10 kHz up and start working stations calling up 4 or 11 kHz, pretty soon your transmitting frequency will be obliterated and your pileup will become unwieldy. Letting a pileup extend over too large a range is also unfair to other stations trying to use the band.

My next piece of advice is for inexperienced DXers who find themselves in rare locations. Amateur Radio is a great way to keep in touch with the world, but if you are in a rare country you are going to attract a lot of attention. That's the way things are. While you might like to ragchew with every station you work, your operating time is going to become more and more aggravating as the pileup turns into a monster. I suggest keeping your contacts brief.

The best way to rid yourself of a pileup is to work everyone in it. Getting mad and leaving the air suits neither your needs nor anyone else's. You may feel that a proper contact requires you to exchange signal report, name and location, but that path is fraught with frustration. Send that info once in a while, and let others decide how much of your time they want to take up in reply. We can look up your name and location online. But what we really want is a confirmed contact, as short as that might be. Unless you are in an extremely rare location, you can probably thin out the pileup in a few days of operating, and then can settle down to a more leisurely pace. Try to establish a tuning pattern and a rhythm. Running a pileup can be a frantic task, but it's also rewarding. If you're trying to kill some time while far from home, there's no better way than running 100 or more stations an hour!

Wayne Mills, N7NG, wrote an informative pamphlet titled *DXpeditioning Basics*, which you can download from **www.arrl.org/awards/dxcc/dx-basics. pdf**.

Chapter 7

Propagation

When I decided several years ago to finish up my first DXCC award, I made a list of countries and entities I hadn't worked, but that had resident radio amateurs. Next, I needed to know when propagation would favor working them, and on what bands. While I was at it, I calculated the times at my location that matched up with weekends at theirs.

At that time in the late 1980s, solar activity was very high. It was high enough that the 6 meter band often opened to Europe. I couldn't operate on 6 meters, but I could operate on 80 through 10 meters, so that's where I focused my attention.

Before sitting down with paper and pencil, I assessed my antenna situation. On 80 and 40 meters my available antennas were too low. Besides, I had learned that if 10 meters was open to a part of the world I wanted to work, I should use that band. Why was that? Because one part of the ionosphere called the D region, attenuates signals at the frequencies we use for DXing. Like the other parts of the ionosphere, D region ionization is greatest on the side of the earth facing the sun. The ionospheric F layer is higher in altitude, and that's where most long distance signals are bent back to earth. F layer ionization is also greatest on the sunlit side of the earth.

So, if solar activity were strong enough to allow the F layer to propagate 10 meter signals and they were least affected by the D region, 10 meters was the place to be. Indeed, on many weekends I could hear and work all parts of the world on 10 meters at the same time. A 100 W station equipped with simple dipole antennas could make DXCC in a day. Long path 10 meter openings to Japan happened almost every morning. Like all good things, good propagation doesn't last forever. As I write this book we're at the opposite end of a solar cycle, and 10 meters is a lonely band.

What's all this talk about the D region? What's the F layer all about? And what's the Sun got to do with it? When you sit down at the radio, you want to know what bands are open, and to what parts of the world. A propagation prediction program will tell you, but it will be more helpful if you know more about how propagation works.

The Ionosphere

Fortunately for all living things, the Earth is shrouded in a gaseous *atmosphere*. The atmosphere is most dense at the Earth's surface. As the distance from the Earth's surface increases, atmospheric density decreases. This is bad for mountaineers; but it's good for DXers.

Our Sun is visible because much of its radiated energy is at visible wavelengths. The Sun also radiates energy at other, invisible wavelengths. One characteristic of electromagnetic radiation is that shorter wavelength radiation has a higher level of energy. X-rays, for example, can damage living tissue. X-rays have much shorter wavelengths (higher frequency) than visible light. X-rays, gamma rays and ultraviolet (UV) radiation all play a role in radio propagation.

These high energy waves (sometimes called *photons*, since electromagnetic radiation has characteristics of both waves and particles) are capable of knocking electrons out of their orbits around atomic nuclei. Every element (iron, helium, oxygen, nitrogen) has a specific number of electrons in orbit around its atoms. When an element loses one or more electrons, it's said to be *ionized*. High energy solar radiation knocks electrons away from atmospheric atoms, resulting in *free electrons*. Nature strives for balance, and those electrons want to *recombine* with ionized atoms, which also want to return to their natural state. Were those free electrons not on the loose, however, long distance ionospheric propagation would be impossible.

The F Layers

The ionosphere was an unknown phenomenon before Marconi and others began communicating via radio over long distances. Scientists assumed that radio waves could not travel much farther than the visible horizon. After several experiments disproved that theory, another explanation was required. The existence of the ionosphere was proved by transmitting short bursts of radio energy straight up and listening for echoes. At different times of the day, and on different frequencies, the time between transmission and reception varied. Scientists now had the clues they needed to understand the solar effect, and also the height of the ionized areas they were discovering. In time, we learned that the atmosphere at about 150 miles (250 km) above the surface was just dense enough that solar radiation would ionize it enough to bend radio signals at the frequencies we use for DXing.

Scientists correctly assumed there could be more than one highly ionized region in our upper atmosphere, so they named the first one they discovered the *F layer*, also known as the *F region*. As luck would have it, during the daytime there are often two F regions, called F_1 and F_2, but the pioneering scientists had the right basic idea. What happens when a radio signal strikes the F layer?

We sometimes refer to *bouncing* a signal off the ionosphere, like bouncing a ball off a wall. Waves don't exactly bounce like balls though. When a wave enters an ionized region it transfers some of its energy to the free electrons there. So far as the wave is concerned, it is entering a medium of greater density, like light waves passing from air into water. The wave is slowed slightly below the speed of light, depending on the density of the medium it has entered. This slowing down increases the deeper the wave moves into an increasingly ionized layer above the Earth. The wave's energy is passed from electron to electron and the path of the wave is *bent* back toward the Earth.

All radio waves passing through an ionized region are thus *refracted*, a fancy term for being bent. If they are refracted enough, the wave passes back out the bottom of the region, and are re-radiated by the free electrons toward the Earth's surface. If the ionospheric region is insufficiently ionized, the incoming wave plows through it and then continues out into outer space. As the wave moves past electron to electron, some of its energy is lost in the form of heat.

How much the wave is bent depends on how many free electrons exist in the region (that is, the degree of ionization) and the wave frequency. The angle at which the wave strikes the region also affects how much it is refracted. A radio wave at 10.1 MHz striking the F layer at a shallow angle may be bent back to Earth. At steeper angles, waves of the same frequency will not be bent enough to make it back to Earth.

As ionization level varies throughout the day, a radio wave that is easily returned to Earth at noon may disappear into space a few hours later. To know if we can reliably communicate between two points by means of the ionosphere, we have to know the level of ionization, the frequency and the angle at which the wave strikes the ionosphere. We refer to the last parameter as the *angle of radiation*. We'll look more closely at radiation angles in Chapter 8, Real-World Antennas.

F Layer Calculations

The ionosphere has been studied in depth since the 1920s. We now have a great deal of statistical information about it. That allows us to make accurate predictions, *under normal circumstances*. It so happens that solar radiation (*solar flux*) at a wavelength of 10.7 cm (2800 MHz) correlates well with empirically determined F layer ionization. If we know the 10.7 cm solar flux we can calculate the frequencies at which the F layer will propagate radio waves.

As mentioned earlier, during daylight hours the F layer can separate into two distinct parts, called F_1 and F_2. At night, the free electrons, no longer bombarded with solar radiation, find their way back to atmospheric molecules and the two layers merge into one.

Sunspots

Sunspots are a favorite topic of discussion among DXers. Sunspots are areas of extremely high magnetic fields on the sun's surface. Though still very hot, they are cooler than the surrounding visible areas, and thus appear to be darker than the rest of the Sun. They are visible *when viewed safely through dense filters*. We care about sunspots because they grow and decline in number at a predictable rate. When they are numerous, solar UV radiation is at its highest and so is F layer ionization. The number of sunspots varies over an approximately 11 year period, which we call the *Sunspot Cycle*.

Even when sunspots disappear for a while, the sun still emits enough UV radiation to allow some propagation on Amateur Radio bands of interest to DXers. During peak years of solar activity DXers drink a lot of coffee, because the bands never seem to close down. The 20 meter band may stay open all night and there are breathtaking openings on 10 and 6 meters. When the sunspots go away, savvy DXers move to the lower frequencies — and usually still find plenty of exciting DX to work.

Our Enemy, the D Region

What is the D region, also known as the D layer? (I'll mention the E layer further on.) The D region is the part of the ionosphere closest to Earth. The D region is of interest because of its *negative effects* on propagation. Because the atmosphere is more dense nearer the surface, when the D region is ionized, it greatly attenuates frequencies of interest to DXers, particularly the lower frequencies. As you move higher in frequency the amount of attenuation decreases. Depending on your power level and the angle of radiation, radio signals can still penetrate the D region and continue on to the F layers to be propagated over long distances, but they lose some of their strength doing so. Below a certain frequency, too much of the signal may be absorbed by the D region to permit any F layer propagation. That is why signals on 80 or even worse 160 meters are much weaker during the hours of sunlight. The D region soaks them up.

Fortunately, free electrons in the D region recombine with ionized molecules very quickly after sunset, and thus D region absorption diminishes rapidly after nightfall. Now D region absorption never goes away completely, but as sunset approaches, frequencies below the 40 meter band suddenly spring to life.

Maximum and Minimum Usable Frequencies

The F layers and D region together establish the highest and lowest frequencies we can use at a particular time. D region ionization also depends on solar activity, though not to the extent seen in the F layers. Let's say it's 11 o'clock in the morning at my East Coast location in Florida (1500 UTC). The D region all around me is heavily ionized, so I have to use a frequency high enough to penetrate it.

Consulting the experts, I learn that frequencies above 12 MHz will result in an adequate signal level on the other side of the D region. That's the *minimum usable frequency*, also called the *lowest usable frequency* (LUF). Knowing that higher frequencies will be attenuated even less, I head for 10 meters (28 MHz), looking for some DX. But the band is dead! Unfortunately, the experts forgot to tell me that the F layers aren't ionized enough to propagate 10 meter signals to anywhere. And I forgot to tell them where on the Earth I want to communicate. Whoops.

Well, let's start with Europe. It's just an ocean away and plenty of radio amateurs live there. At this point in the sunspot cycle, at this time of day, the F layers will not propagate signals to Europe at frequencies higher than 18 MHz. For that path, at this time, 18 MHz is the *maximum usable frequency* (MUF). If I want to work some Europeans right now, I'll have to use either 20 or 17 meters. Even then, the F layers are so lightly ionized I can't expect great signal reports.

What about later tonight, when the D region has mostly dissipated? After dinner, let's say 8 pm local time, which is 0000 UTC, where can I hope to find Europeans? Yes, the D region is no longer limiting the minimum usable frequency, but the F layer has also thinned out. Now the MUF is about 9 MHz. If I want to work Europeans tonight I'll have to be on 160, 80 or 40 meters. (The 160 meter band has its own set of propagation anomalies, discussed below.)

However, what a difference a few years can make. When sunspot numbers are higher, I'll still be able to work Europe on 160, 80 and 40 meters at night, but I'll also be able to work them on 20 meters. Since D region ionization always exists, 20 may be a better choice than the lower (in frequency, not wavelength) bands. And 10 meters will be open in the mornings.

Again, D region absorption decreases as you go higher in frequency. At the same time I'm struggling to work weak Europe on 17 meter now during a sunspot minimum, signals will be booming in on 10 meters when solar activity increases.

Multihop Propagation

The height of the F layers sets an upper limit to the geometry of how far a wave can travel before striking the Earth. Even if the wave enters an ionospheric layer at a very shallow angle (a low angle of radiation), the longest distance traveled in a single *hop* from Earth-to-ionosphere-to-Earth is about 2500 miles (4000 km). To cover longer distances the signal requires multiple hops from the Earth back up to the ionosphere and then back down to Earth again. From my location in the Southeast US, the signal must enter and leave the F layer at least twice to travel to Europe, which is 4200 miles away.

Waves expand as they move through space (think of ripples on water), and the power level at some distant point is proportional to the *square* of the distance. At twice the distance the signal is only one quarter as strong; at four times the distance the signal is 16 times weaker. We know some of our signal's energy is dissipated in the ionosphere, and even more is lost when it is reflected from the Earth's surface. Antennas that radiate at lower angles help minimize the number of hops, but multihop propagation is a fact of life. Longer paths may require six hops, though a lower angle of radiation may require only five hops. As the height of the F layers varies throughout the day, so does the number of hops for a given angle of radiation. We'll look more closely at these factors in Chapter 8.

Long Path Propagation

The *long path* is opposite to the shortest arc between two points on the globular Earth. If the shortest path to another station is at a compass bearing of 45°, the long path bearing is 180° away, or 225°. Both the D region and the F layers must cooperate to provide long path propagation. Let's look at the path between my Florida location and Perth, Western Australia (VK6). The short path distance is 11,400 miles (18,400 km), while the long path distance is 13,500 miles (21,700 km). An hour before my sunset in the Northern Hemisphere midwinter (2130 UTC, December 21), the short path with a signal launched westward at 265° is completely in daylight, since it is 15 minutes after sunrise in Perth.

During periods of low solar activity, the F layer cannot reliably support short path propagation on higher frequencies (ones that aren't absorbed by the D region). The MUF is 14.5 MHz, but signal levels on 20 meters will be low, because eight lossy hops are required to span the short path on 20 meters at 2130 UTC, and F layer ionization is low. There aren't enough free electrons in the F layer to transfer sufficient energy at 14 MHz.

Except close to either end of the circuit, however, the long path, with a signal launched at 80°, is completely in darkness. Signals on 40 meters can make it through a D region that is just in sunlight with relatively small attenuation, and at the same time there are enough free electrons in the F layer between here and Perth to propagate signals at 7 MHz. The long path still requires many ionosphere-to-Earth hops, but signal levels will be strong enough for reliable communication on 40 meters, and often on 30 meters as well.

Sometimes what seems like long path propagation really isn't. I often see DX spots proclaiming "LP" for long path, that aren't. In the example above, if I were working into VK6 on 20 meters, it would probably be on the short path. At the same time and date but on lower frequencies, it would almost certainly be via the long path.

Skewed Path Propagation

The height of the F layers in the Earth's ionosphere depends on the strength of solar radiation. Areas of the ionosphere that face the sun are more highly ionized and are at higher altitudes than other areas. So the F layers are not perfectly spherical around the globe, nor are they uniformly ionized. To a radio wave launched from the Earth the ionosphere does not look always look like a flat plane. It may instead be tilted, toward or away from the wave, or sideways to it.

We know that radio waves expand as they travel, forming a wave front, rather than a pinpoint. When a radio wave moving in a certain direction enters the ionosphere, it may not be refracted along the same bearing. Under certain circumstances, radio waves may travel between two points on very indirect paths. For example, when the ionosphere above the equator is more affected by solar radiation than it is at higher latitudes, signals between Europe and North America will travel much longer distances than expected. Radio amateurs using directional antennas may find they have to point them way off the expected direct headings for best results.

Such *signal skewing* affects long path propagation as well, sometimes resulting in completely unexpected openings. That signal you're hearing may not be a bootlegger after all. Never mind what your intuition and your propagation prediction software are telling you. Work the station and then try to understand how the contact came about. (DXers call this *WFWL*, standing for "Work first; worry later.")

The E Region

As you would expect, the E region lives between the F region and the D region. The E region plays an important role in long distance communication, in often unexpected ways. A signal traveling down from the F region may encounter an area of E region ionization that refracts the signal upward again. The signal travels in a *duct* defined by the F and E layers, and absorption is usually much less than if the signal were reflected from the lossy Earth instead.

Another anomaly that occurs at about the same altitude as the ionospheric E layer is *sporadic E* propagation, abbreviated E_s. Sporadic E propagation is ionospheric, but the source of ionization may not be solar. Small clouds of ionization at the height of the E layer can appear above thunderstorms, and may be caused by wind shear, ripping apart areas of the upper atmosphere. Sporadic E propagation is intense, capable of refracting signals into the hundreds of MHz. The clouds are usually not very large, and they move quickly, sometimes disappearing, only to be replaced by other clouds. While Sporadic E propagation is of most interest to VHF enthusiasts, it frequently creates exciting openings on 10 and even 15 meters as well. During times of high solar activity, when 10 meter F layer propagation is commonplace, Sporadic E clouds can help get your signal into the F layer at lower latitudes, often producing unexpected band openings.

Grayline Propagation

The *grayline* is an area straddling the terminator between daylight and darkness. Two points along the grayline are both either near sunrise or sunset. Because the D region is closer to the Earth's surface, it is ionized later than the higher F layers, and its ionization dissipates earlier at sunset. If you are listening to a station to the east of you, you may notice a large increase in signal strength about the time of that station's sunrise. The same effect occurs in reverse at sunset. Two stations located along the grayline will both experience this effect at about the same time. Grayline signal enhancement may occur on either short or long paths. Because the grayline is thin and moves rapidly westward, grayline openings may not last more than a few minutes.

One Way Propagation

If you are in the northern half of North America, you may find your signal refracts from a Sporadic E cloud, back to earth, and then into the F layers at a point where the ionosphere is capable of refracting them again. Signals coming from the opposite direction may not refract along the same path, missing the Sporadic E cloud, for example. Thus, you may be heard by, or may hear, stations you cannot work.

Similar effects sometimes occur when signals are ducted between the F and E regions. Due to ducting, the last leg of your signal's path into a distant location may be at an angle that allows the other station to hear you. Unless that station can radiate sufficient power at the same angle, however, the return signal may take a completely different path, perhaps going off into space instead. One way propagation is unusual, but it happens often enough that you should do some investigation before tearing apart your antenna system and amplifier, and tearing out your hair!

Geomagnetic Activity

Our sun does not act alone in the drama of radio propagation. Free electrons produced by solar radiation are strongly affected by Earth's magnetic field. At the same time, the geomagnetic field can be bent and shifted by particle eruptions from the Sun.

These eruptions are called *solar flares* and *coronal mass ejections* (CME). A larger eruption can send a Mt Everest sized mass of hot plasma hurtling into space. If the Earth happens to lie in the path of this cloud, the effects on our ionosphere and our geomagnetic field can be severe.

The geomagnetic field protects us from most dangerous radiation by deflecting it to the polar regions. The geomagnetic field is displaced by the radiation, forming a tail on the dark side of our planet, and pulling the polar auroral regions down toward that side. The amount of displacement shifts according to pressure from the *solar wind* and particle storms launched from the Sun. As the field shifts, magnetic lines of force are pushed and bent. Because the ionosphere consists of free electrons, which are affected by magnetic fields, the F and E layers may be ripped into shreds. At the same time, higher energy particles can increase D region absorption. As if that weren't bad enough, the concentration of energy where the magnetic lines of force converge at the poles greatly increases absorption of signals traversing those areas.

Rapidly shifting magnetic fields induce enormous dc currents in power lines as well, often causing widespread power outages, especially in areas nearer the poles. **Fig 7-1**, obtained from **www.sec.noaa.gov/pmap/pmapN. html** (use **www.sec.noaa.gov/pmap/pmapS.html** for the South Polar region

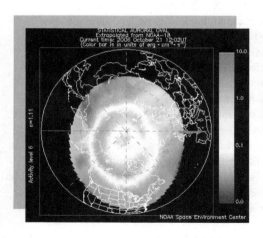

Fig 7-1 — Representation of the northern auroral oval developed from satellite data. As solar particles increase the intensity of an aurora, the area within the auroral zone extends farther from the pole. Solar wind pushes the auroral zone farther from the pole on the dark side of the earth.

image) about two hours before sunrise in my location. This is a time when the 160 to 30 meter bands may be open from my location to the Far East. Because the aurora extends so far south, though, signals may be attenuated to the point of inaudibility. (It never hurts to turn on the radio and listen, however!)

Geomagnetic storms can cause propagation blackouts, which often arise in a matter of minutes. Scientists call the effect a *sudden ionospheric disturbance* (SID). Within a couple of minutes, signals drop into the noise, and the bands are dead for several hours. The effect can make you think your receiver is broken or that your antenna fell down. Signals will gradually reappear several hours later. Fortunately, we have some ability to predict these outages from satellites in space, though, like the weather, all we can do is talk about them.

Intense geomagnetic storms usually increase the visible aurora, which is good for VHF operators living near the poles. The aurora can become so intensely ionized that it reflects signals as high as 432 MHz. Because the magnetic field is in rapid motion, signals are modulated by the Doppler effect, making CW and SSB signals sound ghostly, even ethereal. Operators on both ends of a path point their antennas at the aurora, rather than at each other. VHF DXers in North America probably won't get outside the continent during an auroral opening, but in Europe it's possible to work many other countries.

Propagation on 160 Meters

The 160 meter band is the most challenging of all bands commonly used for DXing. Because of its long wavelength, antennas are physically large. But propagation also presents its own set of challenges. Most propagation prediction programs do not calculate for frequencies below 3 MHz. While 160 meter signals are still propagated by the ionosphere, there are various propagation modes and effects we only partly understand.

For example, experienced 160 meter DXers often report a *spotlight* effect, where distant signals are very strong in one location, yet weak or inaudible a few dozen miles away. Another effect often noted on 160 is enhanced propagation just prior to the onset of a geomagnetic storm.

Because of its low frequencies, 160 meters is always more affected more drastically by D region absorption than the higher frequency bands. This has led many to assume propagation on this band is better during periods of low solar activity. While low sunspot numbers usually drive more people to *Topband*, as 160 meters is often called, propagation there is often better during times of higher solar activity. While it's unlikely you'll work a country on 160 meters that you couldn't work on another band, DXCC on this band is possible from even a modest station.

A good introduction to 160 meter propagation is "The 160-Meter Band: An Enigma Shrouded in Mystery," by Cary Oler and Dr Theodore J. Cohen, N4XX. Originally published in the March and April 1998 issues of *CQ*, the article is available for download from **www.spacew.com/cq/cqmar98.pdf**.

I mentioned that 160 meters is the most challenging DX band, and sometimes we can use some help and encouragement! If 160 piques your interest, I recommend two books published by ARRL: *Lowband DXing*, by John Devoldere, ON4UN, is the classic text for 160 meter DXers, and it also contains a wealth of information of interest to 80/75 and 40 meter DXers as well. *DXing on the Edge — The Thrill of 160 Meters*, by Jeff Briggs, K1ZM, is aptly titled. Filled with historical lore and practical tips, it will surely inspire you to give 160 meters a try. The two books perfectly complement each other.

Propagation on 80/75 and 40 Meters

DX propagation on these bands is possible only through areas of darkness. Both ends of a path may be near either side of sunrise or sunset if the angles of radiation are low. Because the F region is higher in altitude than the D region, the F region is ionized earlier and stays ionized later than the D region. For up to about an hour after your sunrise, when the D region is not yet fully ionized (and hence lossy), you will find good propagation to the west. Similarly, prior to your sunset, D region ionization declines rapidly, and propagation to the east is possible while you are still in daylight.

The 80/75 meter band is often open all night even during years of low solar activity, and so is 40 meters, though not as often. Better conditions usually follow the sun. As the sun's illumination moves westward, signals at locations that are near sunrise are often enhanced. Unfortunately, good conditions are no guarantee anyone is awake and on the air in those locations. From the continental US, Thursday nights through early Friday mornings are a good time to look for the Far East, since it is Friday night over there.

Don't overlook long path propagation on these bands, which happens more often than many DXers realize. Starting an hour or so before your sunset, tune the bands for weak signals, though long path signals are often stronger than you might expect.

Propagation on 30 Meters

The 30 meter band has properties resembling those on lower and higher frequencies, and can provide DX openings day and night throughout the sunspot cycle. Signals are often enhanced at sunrise and sunset times at either end of a path and long path openings are common.

Propagation on 20 and 17 Meters

Twenty meters is the favorite band of most DXers. If a DX station shows up anywhere, it's likely to be on 20 meters. During years of high solar activity, 20 and 17 meters are often open all night, causing much excitement and loss of sleep for DXers. When the sunspots are few in number, 20 will usually open every day, but openings are short and signals are sometimes weak. The same conditions prevail on 17 meters. Because of lower activity there, it's often easier to work DX on that band than on 20, even though signal levels are often about the same. In times of higher solar activity, 17 meters is a better bet due to lower D region absorption.

Propagation on 15 and 12 Meters

These two bands are most useful when solar activity is high, though they will often open for a few hours each day during low activity years as well. Thanks to lower D region absorption, if these bands are open, signal levels are often much stronger than on 20 and 17 meters. For this reason, always listen on the highest band that is open. These bands tend to close down over darkness paths, unlike 20 meters, which can stay open all night, even over darkness paths.

Propagation on 10 Meters

The 10 meter band has propagation conditions resembling those of 15 and 12 meters, but often provides openings due to sporadic E propagation. During periods of high solar activity, 10 meters can be open to all parts of the world at the same time, and long path openings can be positively breathtaking. F layer 10 meter propagation lies along daylight paths, but D region absorption is very low. For example, from the Northeast US in peak solar activity years, the long path to the Far East will open every morning, since it is entirely in daylight. If you have to get to work on time, I don't recommend turning on the radio in the morning!

Propagation on 6 Meters

Aptly called *The Magic Band* by its many fans, probably no other Amateur Radio band offers the variety of propagation modes you'll find on 6 meters. Even during years of low solar activity, contacts between Europe and the Eastern and Central United States are possible. Thanks to reduced D region absorption and the relatively short wavelength, 100 W stations with small antennas can do very well.

Other propagation modes found on 6 meters include tropospheric refraction, sporadic E, meteor scatter and auroral propagation. These modes don't usually provide openings usable for really long distance exotic DXing from North America. If your transceiver works on 6 meters it's worth monitoring the band, especially during VHF contests.

Summing Up

As we've seen, the bands above 10 MHz are most useful during daylight, along daylight paths. They are also more affected by solar activity. The bands below 10 MHz favor dark paths. The 30 meter band is interesting because it shares characteristics with both those above and below it in frequency. During years of high solar activity, the 20 and 17 meter bands may stay open all night. When solar activity is low, the MUF may drop below 3.5 MHz overnight, effectively closing down even the 80 and 75 meter bands.

Propagation Prediction

Propagation prediction is never 100% certain. One reason why many commercial and military communications networks have left the HF bands is the unreliability of ionospheric propagation. On the other hand, certainty works both ways. A propagation prediction program may suggest a possible band opening that doesn't occur. At other times you will discover band openings not predicted or expected. Use propagation programs as a guide, but spend as much time listening as you can spare.

Prediction programs rely on several key pieces of information. The data are available online in the Geophysical Alert Message at **www.sec.noaa. gov/ftpdir/latest/wwv.txt** or off the air from WWV and WWVH. These parameters are: Solar flux (related to Smoothed Sunspot Number), and the A and K indices. Later in this section I provide information on two propagation prediction programs available for free download. There are many others, and everyone seems to have a favorite. To help you learn more about radio propagation, ARRL has compiled dozens of articles and Web links, at **www. arrl.org/tis/info/propagation.html**. (Some, but not all, are available only to ARRL members.)

Solar Flux and Sunspot Numbers

Solar flux represents a measure of the ionizing effect of solar radiation. Flux is measured at 2800 MHz, though radiation at that frequency has no known effect on ionization. That frequency was chosen because it will almost always penetrate the atmosphere for reception by earthbound receivers. As it happens, there is a correlation between the strength of solar radiation at that frequency and the ionizing effect of solar radiation.

Scientists have developed an equation that relates the number of sunspots and solar flux. Some programs will accept as inputs either sunspot numbers or solar flux. Make sure you choose the right designator. The number of sunspots can vary throughout the day, so a Smoothed Sunspot Number (SSN), averaged over a 12 month period, is sometimes used. Because the ionosphere does not respond quickly to changes in sunspot numbers, running predictions based on current solar flux measurements is less accurate than using the monthly smoothed sunspot numbers or smoothed solar flux. This information is also available from ARRL in its weekly propagation bulletin. This bulletin is transmitted on W1AW, and available online at **www.arrl.org/w1aw/prop/**.

During periods of high solar activity, the daily sunspot number and solar flux may vary widely from day to day, so use the weekly mean numbers instead. For example, in the last week of December 2000, when solar activity was high, daily sunspot numbers varied from 130 to 189. In late July 2006, a time of low solar activity, daily sunspot numbers varied from 11 to 26. ARRL propagation bulletins now give daily and weekly mean solar flux values as well.

A and K Indices

The *A* and *K Indices* represent the level of geomagnetic disturbance. Higher numbers indicate greater disturbances, which negatively affects ionospheric propagation. K indices are measured at three hour intervals. The A Index summarizes K indices over a 24 hour period. The K Index is measured with a magnetometer, which sensitively detects shifts in the geomagnetic field. Since the geomagnetic field varies around the planet, a measurement taken in one location does not necessarily reflect conditions elsewhere. This is especially important when considering polar paths.

You should note that the WWV forecast is based on magnetometer readings taken in Boulder, CO, in the Western United States, at a latitude of 49 degrees North. DXers who use the 160, 80 and 40 meter bands should also consider A-indices measured at more northerly sites. This information is also available online, at **www.sec.noaa.gov/ftpdir/latest/AK.txt**.

The *Planetary K Index* is calculated as the standardized mean of K indices measured at 13 sites around the world. This information is graphically presented at **www.sec.noaa.gov/rt_plots/kp_3d.html** for a three day period. If you're interested in general propagation conditions, the *Planetary A Index* is the one to use. K Indices can vary from 0 to 9. Indices greater than 4 indicate disturbed conditions, when propagation conditions are generally poor. East-west paths are most affected by geomagnetic disturbances, but north-south paths are also attenuated. When the K Index is greater than 4 you're more likely to find signals to the north or south of you, but they will be weaker and subject to deeper fading.

While propagation programs take geomagnetic variations into account, the global differences may cause unpredictable effects. One tool I've found helpful for predicting polar path openings is the NOAA POES (National Oceanographic and Atmospheric Administration Polar Operational Environmental Satellite) images, shown in Fig 7-1 and discussed earlier in this chapter. The resolution of the auroral ovals could be better, but with experience you'll be able to tell when propagation paths passing near or through them should be usable. For example, from my location in the

Southeast United States, if southern Alaska is covered by a yellow or red area, I won't expect good propagation on 160 to 40 meters to the Far East, a northwesterly path crossing that region. I also won't expect good propagation on those bands to Central Asia. If the solar flux is high enough to suggest openings on 20 meters and the higher frequencies, I'll expect to hear fluttery signals from those areas, since the signals are modulated by the shifting aurora.

I've also found that signals from my location to Great Britain and farther north on the lower frequency bands are greatly attenuated, even when the British Isles are covered by a white oval. I'll hear signals from France, only a little south, but not much from stations farther north. By the way, you'll notice that the auroral oval extends farther south on the dark side, an indication of how much the geomagnetic field is stretched by solar radiation.

Basic Propagation Predictions: **W6ELProp**

The propagation program I use most often is *W6ELProp*, written by Sheldon Shallon, W6EL. It's available for free download from many sources, which you can find with any search engine. *W6ELProp* works from a database. Although it has been around for a long time, I still find it very useful. As it is simple to operate, I'll use it as a basic example.

Before using the program, I have to enter my latitude and longitude. Unless I move, I only have to do this once. Then I select the bands for which I want predictions. In the days of slow PC processors this choice greatly affected the time it took to get results. With a newer computer system the program can output results for all bands from 3.5 to 30 MHz in a few seconds. This program also asks for information about your station power and antennas. Under marginal propagation conditions, a station with better antennas, running more power, will be able to hear and work more stations. Unless you have an exceptional station, the program defaults will be good enough to get started.

Now I'm ready to get some predictions. Most of the time I'm more interested in parts of the world, rather than specific locations. The exception is when a DXpedition is pending and I want to know the best times to work it. I've been concentrating on 80 and 40 meters lately, but I tell the program to make predictions for 30 meter as well.

I choose a DXCC entity prefix for the part of the world I'm hunting. In *W6ELProp* I can enter the prefix directly or choose it from a list. (You can manually update the program's prefix database as needed.) Let's say I'm after the Middle East. I'll enter "4X," the most common prefix for Israel. (I can also manually enter the latitude and longitude for the station, if I know it. This is handy for determining sunrise and sunset times when propagation may be enhanced.) The last bit of information I have to provide is the solar flux and the A-Index. Now I'm ready to click "OK" and see how things look.

Prediction Tables

W6ELProp produces a table in 30 minute increments covering a 24 hour period, based on information in its database. The information includes not only the sunrise and sunset times on both ends of the path, but the Maximum Usable Frequency for each time period, and the expected signal strength. An added feature is a letter code (A to D) relating to the probability that signal strengths will reach the predicted level. **Fig 7-2** shows predictions for a two hour period on November 1 for a period of low solar activity.

UTC	MUF	3.5 MHz	7.0 MHz	10.1 MHz
0000	10.0	40 A	32 A	29 C
0030	9.7	37 A	32 A	29 C
0100	9.4	37 A	32 A	29 C
0130	9.2	37 A	32 A	29 C
0200	9.0	37 A	32 A	29 C

Fig 7-2 — **W6ELProp** predictions for a two hour period for the 80/75, 40 and 30 meter bands. See text for details.

From this information I can see that my best chances of working into the region during this time period are on 80 or 75 meters. Assuming similar antenna and transmitted power situations, signals on 40 meters would be 8 dB weaker. Of course, there's no guarantee that anyone in the Middle East will be on the air between 0000 and 0200 UTC. That time period is shortly after sunset there, and they may be having dinner.

The default for W6ELProp predictions is the short path between two locations. Sometimes the long path may be better, however, and sometimes that's the best way to work stations on the far side of the globe from you. W6ELProp will predict long path openings too, with a single mouse click. I find the long path predictions less reliable, but still useful. During some times of the year, some parts of the Far East can only be worked from my location via long path. While the program doesn't always predict long path openings that I have observed, the mapping feature has been a great help to me in discovering them.

If I'm planning a night of DXing on 80 and 40 meters, I know that almost every part of the world will be available at some time. If I want to know more I can print out a listing of these forecasts for up to 15 distant locations at a time. This kind of information is most useful to me before a contest. It helps me decide what times I want to be on either band, and also helps me spot times when I may want to grab a nap.

Mapping Features

W6ELProp also has three mapping features. The one I use most often is the Great Circle Map (**Fig 7-3**). Here I can see what other parts of the world are in darkness and hence might be audible here on 80 and 40 meters, how much of the path is over land and to get a rough idea of how long those bands may be open to the parts of the world I'd like to work. This same information can also be displayed in rectangular format. **Fig 7-4** shows a portion of the rectangular map generated by *W6ELProp* for the same path shown in Fig 7-3.

It's on long path openings that the maps really pay off. On my favorite bands, 80 and 40 meters, both ends of the contact must be either in darkness or close to either sunrise or sunset. Also, the path between us must be in darkness or fall along the sunrise/sunset terminator, the grayline. Let's have a look at the path between my station and YB1A near Jakarta in Indonesia. I worked YB1A by long path on March 6 at 2259 UTC. **Fig 7-5** shows the path between us on that date and time. The short path goes up and to the left, northwest from me, and travels mainly over sunlit geography. The long path goes to the southwest, down and to the right, traveling mainly a dark path over South America. You can see that while it was almost

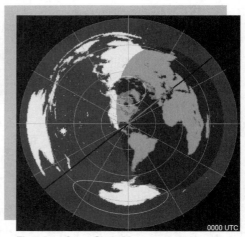

Fig 7-3 — Great Circle Map generated by **W6ELProp**. The map depicts areas of daylight and darkness, and the long and short paths between my Florida location and the Middle East.

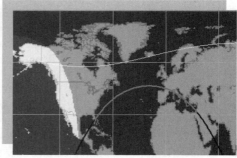

Fig 7-4 — Part of the rectangular map generated by **W6ELProp** for the same path as shown in Fig 7-3.

sunset on my end, it was right at sunrise in Indonesia.

Now look at the path for the same time, but on November 1 (**Fig 7-6**). Neither a 40 meter short path nor a long path to YB is likely on November 1 at sunset in Florida on that day, since YB1A would be in full sunlight.

Changing the times (or using the prediction part of the program) let's me see that there will be short path propagation around the time of my sunrise on March 6 (**Fig 7-7**). The predicted short path signal strengths are about the same for November 1 or March 6 at Florida sunrise. From the map, I can determine that there will be no long path openings to Indonesia on March 6 at my sunrise because much of the 40 meter path over South America from Florida is fully illuminated by the Sun.

On March 6, YB1A's long path signal was much stronger than predicted. Had I relied on the prediction tables alone I might not have caught the opening. Like weather forecasts, propagation prediction programs come with no guarantees. When combined with operating experience,

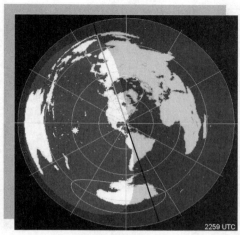

Fig 7-5 — Great Circle Path between KR1S and YB1A at 2259 UTC on March 6. The long path is in darkness and opened, although the program did not predict the opening.

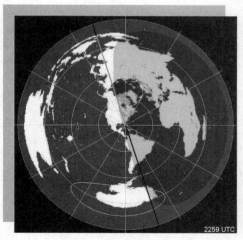

Fig 7-6 — Great Circle Path between KR1S and YB1A at 2259 UTC on November 1. Because both long and short paths are in daylight, contacts on 160 to 30 meters would not be possible.

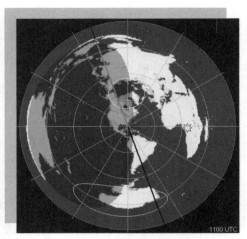

Fig 7-7 — Great Circle Path between KR1S and YB1A at sunrise in Florida, 1100 UTC on November 1. At this time, the short path is entirely in darkness, and contacts on 160 to 30 meters *may* be possible.

though, they can give you insights that lead to exciting and unexpected contacts.

ARRL has provided a tutorial for *W6ELProp*, written by Carl Luetzelschwab, K9LA, available to anyone for free download, at **www.arrl.org/ tis/info/pdf/W6elprop.pdf**.

Advanced Propagation Predictions: *VOACAP*

VOACAP is a program developed over about 30 years by several US government agencies. *VOACAP* provides a wide range of graphical and text data, though the sheer volume of data can be overwhelming. If you're willing to invest some time to learn its features though, it is an excellent educational tool. You can download *VOACAP* for free from **www.voacap.com/**.

VOACAP is also useful for predicting openings to specific areas over a period of time, during a contest, for example. An online paper by R. Dean Straw, N6BV, shows how he did this for a domestic contest, the ARRL Sweepstakes. The principles can be applied to any contest, however, and are worth studying. You can read or download the paper at **www.voacap.com/ documents/N6BV_Visalia_2006.pdf**.

ARRL has provided a tutorial for *VOACAP*, written by K9LA, available to anyone for free download, at **www.arrl.org/tis/info/pdf/Voacap.pdf**.

Further Reading

The ARRL *Antenna Book* covers radio propagation in some detail. A comprehensive discussion of ionospheric radio propagation is provided by *The NEW Shortwave Propagation Handbook*, by George Jacobs, W3ASK, Theodore J Cohen, N4XX, and Robert B Rose, K6GKU. (Hicksville: CQ Communications, Inc.) Other books on propagation are available from the ARRL Bookstore, **www.arrl.org/catalog/**.

Ham Cap, a freeware program by Alex Shovkoplyas, VE3NEA, is a handy interface to *VOACAP*. You can download it at **www.dxatlas.com/**.

Chapter 8

Real-World
Antennas

A mateur Radio in the United States has fallen afoul of restrictive covenants and local ordinances. Radio Amateurs often have to make do with smaller, less-effective antennas. High, directional antennas are a big help, but you can achieve DXCC with simpler antennas.

A Few Words About Safety

In our zeal to put up an antenna, we sometimes overlook important safety requirements. Avoid installing an antenna that *could* contact a power or telecommunications line if either one broke. And never use utility poles for antenna supports.

If your antenna is concealed or disguised, what will happen if someone touches it while you're transmitting? If you are fortunate enough to own a tower, never climb it without appropriate safety equipment. Quick, careless trips up the tower may prove fatal.

Finally, consider the biohazards of RF radiation. Small indoor antennas are appealing but may be hazardous to your health. So too may be outdoor antennas, if their locations place them in close proximity to humans. *The ARRL Antenna Book* covers these and many other topics and should be required reading for all DXers.

Angle Of Radiation

All antennas commonly used for DXing radiate RF in all directions, horizontal and vertical. Some designs maximize radiation in certain directions (horizontal angles) and at certain vertical angles. To propagate a signal between two points, we'd obviously like to put most of the RF energy into the horizontal angle or bearing that corresponds to the great circle path between those points. But because the height of the ionosphere is not always the same along the path, and varies throughout the day, we must also consider the vertical angle of radiation, often called the *launch angle*.

Many factors affect the vertical angle of radiation (and reception) — too many to cover in this book. In general terms, lower angles of radiation are preferable, most of the time. The ionosphere can refract signals entering at shallower angles at higher frequencies than will be refracted at steeper launch angles. High angle signals may pass completely through the ionosphere and out into space. Because the ionosphere is constantly in motion, higher-angle signals may propagate better over some paths on some bands at certain times of the day. If you can have only one antenna per band, one that radiates significantly at lower angles (14° or less) is what you want.

Size Matters

To be naturally resonant at a certain frequency, an antenna must be ½ λ long. By means of *antenna tuners* (more properly called *impedance matching networks*) radio amateurs have managed to resonate almost anything imaginable, and they've even worked some DX in the process. Antenna tuners are useful, but are no substitute for resonant antennas.

I'm especially wary of so called *compact transmitting loop antennas.* While only a few feet (about one meter) in diameter, they can often be resonated at frequencies as low as 3.5 MHz. What's the catch? Simply put, the RF current flowing in these loops is extremely high. Even a tiny amount of resistance in the loop causes high losses, meaning you're losing a great deal of power in heat, not radiated RF. The percentage of efficiency is in the single digits.

Another drawback to small transmitting loops is their narrow bandwidth. As you tune around the band you have to keep readjusting the loop's tuning capacitor. Although the radiation efficiency of small transmitting loops is very low, the radiated power may still pose a biological hazard if they are located close to the operator or to his family. Don't use one where humans can come in close proximity to it.

Multiband Antennas

Multiband antennas come in several varieties. Multiband rotary beam antennas use tuned circuits called *traps* to isolate one or more segments of the elements on different bands. See **Fig 8-1**, which shows a typical triband Yagi for 20, 15 and 10 meters. The same trap technique is commonly used in multiband vertical antennas and for some dipoles. Because the traps act as switches, connecting more or less of the overall length or height of the antenna elements, trapped antennas are resonant on the bands for which they are designed. The individual traps introduce losses but allow the antenna to be fed

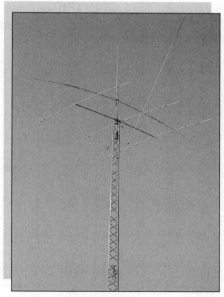

Fig 8-1 — A typical 3-element triband (20, 15 and 10 meters) Yagi at the top of a tower, with a 40 meter rotatable dipole and a VHF vertical antenna above the Yagi.

with coaxial cable, and they do not require the use of an antenna tuner.

Some horizontal wire antennas are designed for use with high-impedance open-wire transmission lines, commonly called *twinlead*. The most popular variation is called the *G5RV*, since it's based on a design published by Louis Varney, G5RV, in 1966. The original design was fed with a combination of coaxial and twinlead cable, with the latter acting as part of a matching network for 20 meters. For use on multiple bands Varney recommended use of only open-wire feed line and an antenna tuner down in the shack.

The G5RV is 102 feet long, less than ½ λ long on 80 meters. This allows installation on smaller lots, which usually implies installation at lower heights, where DX performance is greatly reduced. The G5RV and its variants are very popular and far be it from me to contradict their fans. If DXCC is your objective, though, there are alternatives you should consider.

Dipole Antennas

Probably the world's most-popular antenna, the dipole, serves by itself and as part of larger antennas as well. See **Fig 8-2**. The nominal impedance of a dipole is 72 Ω, while modern transceivers are designed for use with 50 Ω loads. What happens when you connect a piece of 50 Ω coaxial cable to a resonant dipole? It radiates RF, as it should. Don't let that flickering SWR meter disturb you! Turn off that clattering built-in antenna tuner and go work some DX.

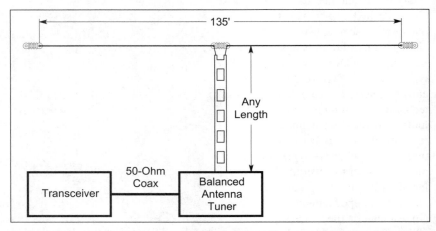

Fig 8-2 — Drawing of a 130 foot long flattop dipole fed with open wire feed line for multiband operation. Many hams feed such a dipole with coaxial cable directly, without an antenna tuner and get very acceptable results on one band. (Courtesy, *Simple and Fun Antennas for Hams*.)

Dipole antennas are noticeably directional when installed at least $\frac{1}{2}$ λ above ground. If you can have only one dipole, try to orient it so it faces the distant place you want most to work. The angle of radiation from a dipole decreases as the antenna height is raised, with $\frac{1}{2}$ λ being about right most of the time. Since the tallest trees most of us are likely to have are about 70 feet (about 20 meters) high, it's hard to install an 80 or 75 meter dipole high enough for effective DXing. If that's the case for you, consider a different antenna for the lower bands and try a dipole on 40 meters.

Because most of the radiation from a dipole occurs near the feed point, it's possible to droop the ends without greatly affecting performance. On frequencies above 10 MHz, large loop antennas may provide better performance than dipoles. And when made into the shape of a vertical triangle they require only one support, not two. I'll discuss more about loops soon.

The Inverted Vee

Inverted vee dipoles are supported at the center, not at the ends. Bending the antenna changes the feed point impedance slightly, such that an inverted vee is usually a little shorter in length than a flat dipole for the same band. I use an inverted-vee on 40 meters, with the center at about 33 feet (10 meters). The included angle between the elements is a little more than 90°, so the ends are about 10 feet (3 meters) above ground. This antenna is less directional than a dipole, and even though the feed point is relatively low it works very well.

Full-Size Transmitting Loops

Single-element loop antennas use elements that are an electrical full wavelength in size. The formula used to calculate length in feet is 1005/F (MHz). Loops are somewhat more directional than dipoles and offer slight gain along their axis. Two loops hung from a single support, 90° opposed from each other and switched with a relay, can be very effective for full coverage of all directions. Loops are less sensitive to proximity to the ground than dipoles, though higher is still better, if possible. I used a dual-loop arrangement on 20 meters for many years. The bottom of the loops was about 10 feet (about 3 meters) above the ground.

Vertical Antennas

Vertical antennas radiate equally in all directions. Unlike even slightly directional antennas, like a horizontal dipole, verticals radiate most of your power in directions you don't want. But even verticals mounted close to the ground radiate more power at the desirable lower angles of radiation.

Half of a vertical antenna is the ground system, usually comprised of *radials*. When a vertical antenna is installed on or near the ground, most of the ground system consists of the earth itself, which is a poor conductor. See **Fig 8-3**. A ground-mounted vertical with a poor ground system has high losses; meaning most of your power is heating the worms, not being radiated. More radial wires, rather than a few longer ones, can have a major impact on performance. Unless you have 60 or more radials, they don't need to be ¼ λ long — ¹/₁₀ λ will be long enough. You can never have too many radials!

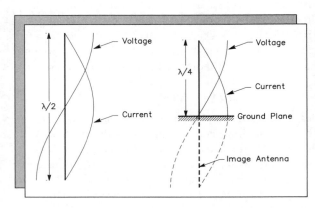

Fig 8-3 — The half wave dipole antenna and its quarter wave vertical ground-plane counterpart. The "missing" quarter wavelength is supplied as an image in "perfect" (that is, high conductivity) ground. (Courtesy, *The ARRL Antenna Book*.)

Elevating the radials helps increase efficiency because it removes the system from the vicinity of lossy earth. But even verticals with elevated radials work better when there is a good radial system on the ground beneath them. With a *perfect* ground system, a vertical antenna has an impedance of about 30 Ω. When connected to a 50 Ω coax the SWR will be about 1.5:1, nothing to worry about. It's unlikely you will have a perfect ground system, however, and ground losses will appear in series with the vertical's intrinsic feed point resistance, raising the overall resistance even closer to 50 Ω.

Vertical antennas for higher frequencies can often be installed ½ λ or more above ground, requiring fewer radials and reducing ground losses. Four radials equally spaced will usually be enough. If they are tilted downwards at about a 45° angle, the antenna feed point resistance will be close to 50 Ω. When I started DXing many years ago, I used a 20 meter vertical on the roof, about 15 feet (4 meters) above ground. It worked, but I always wished for better directivity and gain.

Ground-mounted verticals should be installed as far from other objects, including trees, as possible. Most of the radiation from a vertical antenna is from near the feed point and intervening objects can absorb radiated energy and affect the radiation pattern.

Inverted-L Antennas

The inverted-L antenna is a *bent* vertical. Part of the radiating element length is horizontal and part is vertical. We use inverted-L antennas when we want a vertical radiator but can't support the full height. My 80 meter antenna is an inverted L. The first half, about 33 feet (10 meters) is vertical, and the rest is horizontal. I use four radials that slant upward from the base, supported about 10 feet (about 3 meters) above the ground for the rest of their length. While far from an optimum arrangement, this antenna has performed very well. To work DX you don't have to be the strongest signal on the band, you only have to be readable. Louder is better, of course, but you do the best with what you have.

Rotatable Directional Antennas

Also called *beams*, directional antennas you can turn towards the direction of the other station are commonly used on frequencies above 14 MHz. Some stations even beams on 40 meters, and there are even a few monsters in use on 80 and 75 meters!

The most-common arrangement has a single *driven element*, to which the feed line is connected, and one or more *parasitic elements* behind or in front of the driven element. The elements are usually horizontal aluminum tubes, a design called the *Yagi* after its co-inventor. The driven element is a dipole. Multiband yagis use trapped elements.

Another type of rotatable directional antenna is the cubical quad, or *quad*. Here, the elements are loops. Because the elements are wire, hung on non-conductive spreaders, quads are often lighter in weight than Yagis. They can be very unwieldy to install, however. Both types of beams are subject to wind and ice loads.

A simple three-element rotatable antenna covering the 20/15/10 meter bands is a huge improvement over vertical and dipole antennas. More parasitic elements will provide more gain, but require more support strength and a more powerful rotator to turn them.

Fixed Directional Antennas

The use of parasitic elements to enhance directivity and gain need not be limited to rotatable antennas. If you have enough supports in the right locations you can make a multi-element directional antenna aimed at a particular area of interest. With some switching the antenna can even be made directional in two directions.

Receiving Antennas

Most radio amateurs use the same antenna for transmitting and receiving. On the 160 and 80/75 meter bands, it's possible to make a directional receiving antenna, when a directional transmitting antenna would be impossible due to size constraints. Popular receiving antennas include small loops, the K9AY loop,[1] and the *Beverage* or *wave* antenna. Very small indoor loop antennas are also available, but my experience with them has been unsatisfactory. All antennas work better when they are far from other conducting objects, such as house wiring.

What Do You Need?

You can work DX with almost any antenna, but no part of your station contributes more to success than the antenna system. I have proved this many times over. At various times I have used everything from mobile whips to rain gutters, even a fire escape, and I've worked DX with all of them.

Although there is nothing worse than being able to hear a DX station that cannot hear you, if circumstances permit, it's better to invest in an improved antenna system than in a power amplifier. This is because a better antenna benefits reception as well as transmission. With persistence and skill, you can get on the DXCC Honor Roll using a simple three-element multiband yagi for 20 to 10 meters, plus wire antennas for the lower bands.

If you can't install a beam, don't let that stop you. Antenna experimentation is as old as Amateur Radio. The ARRL Bookstore and the Web are well stocked with antenna ideas you can try at little cost.

---

[1]Gary Breed, K9AY, "The K—A Compact, Directional Receiving Antenna," *QST*, Sep 1997. ARRL members may download a PDF copy of this article at no cost from **www.arrl.org/members-only/tis/info/pdf/9709043.pdf**.

Chapter 9

Contesting
for DXers

F ew subjects are more controversial than Amateur Radio contests. Some non-contesting hams complain "There's a darned contest on every weekend, clogging up our bands." While that is not really true, DX contests can put a lot of signals on the air and that disturbs some people. Contests are, however, an essential part of DXing. And they can be a lot of fun besides.

DXers approach contests from a different perspective than do competitive contesters. Contesters are trying to maximize their scores, while DXers are usually looking for new DXCC entities. DXers could also be looking for entities they haven't worked on a particular band or mode.

QST publishes information about upcoming contests in every issue, including dates and times of operation, the contest exchange (see below) and where to submit your logs. *CQ* magazine also publishes such information for the contests they sponsor.

Anatomy Of A Contest

Amateur Radio contests take place over a predetermined period of time and usually involve only one mode. Some contests are divided into separate operating periods for the different modes. For example, the ARRL International DX Contest is held on two weekends — one for Phone; one for CW.

Contest scores are determined by two factors: The actual number of *contacts* and the number of *multipliers*. The number of contacts part is easy to understand. What constitutes a multiplier depends on the contest. For DX contests, a multiplier is usually an ARRL DXCC entity. One exception is the *CQ* Worldwide DX Contests, in which some multipliers are not DXCC entities.

Contest Multipliers

To obtain the final score, you multiply the number of contacts by the number of multipliers. In most DX contests, multipliers are counted once per band. So if you work the same multiplier once on each of five bands, it counts for five multipliers.

What constitutes a multiplier also depends on the contest. In ARRL International DX Contests, US and Canadian stations try to work as many stations in other countries as possible. Puerto Rico (KP4), the US Virgin Islands (KP2), Alaska (KL7) and Hawaii (KH6) count as DXCC entities for US and Canadian stations. On the other hand, DX stations operating in this particular contest count only US states and Canadian provinces as multipliers.

Contest Categories

Contest entries are divided into categories. They are usually classified by output power, number of operators, and type of assistance used by single operators (such as DX spotting networks). The power classifications are usually QRP (5 W or less output), Low Power (100 W or less output) and High Power (more than 100 W, up to the legal limit of power output).

Some contest stations have more than one operator on the air at one time. These *multioperator* stations often cover all the HF bands simultaneously. Spotting networks help contesters find multipliers, so *assisted* operations enter a different class than unassisted, single-operator stations.

If you plan to submit your score to the contest organizer you must determine and report your classification. If you're only operating in search of DX to work, it doesn't matter, so long as you otherwise comply with the rules.

Contest Exchanges

Contests originally sought to replicate the exchange of messages. Nowadays most contest exchanges are very simple. In the ARRL International DX Contests, for example, US and Canadian stations send a signal report and their state or province. DX stations send a signal report and their output power in watts. The standard signal report in contests is 59 on voice, and 599 (often abbreviated "5NN") on CW and keyboard modes. On CW power levels are also often abbreviated. Instead of "1000" a station might send "KW." The names for states and provinces are abbreviated also on CW.

In the *CQ* Worldwide DX Contests, the exchange is signal report and *CQ* Zone. *CQ* zone numbers are different from IARU zones. Maps and zone tables are available online, and on maps available directly from *CQ*. There are 40 CQ Zones. *CQ* also sponsors two RTTY DX contests, offering mode-flexible DXers even more chances to work new entities. Information about *CQ* contests is available at **www.cq-amateur-radio.com/awards.html**.

One DX contest that is somewhat more challenging is Worked All Europe (WAE). In this contest, European stations try to work stations elsewhere in the world. The exchange is simple enough, signal report and a progressive serial number starting with 001. In a nod to tradition, the WAE contests (there are separate contests for CW and SSB) also features *QTC*, where non-European stations send a message consisting of times, call signs and serial numbers for previous contacts. Each previous contact reported in a QTC gives an additional point to both sender and receiver. Contest logging software, which is absolutely essential to prevent duplicate contacts, will also handle the QTC chores in the WAE contest. Information about the WAE contests is available at **www.waedc.de/**.

For any contests you participate in, by all means use a logging program and make sure it can output a log suitable for submission to the ARRL *Logbook of The World* (LoTW, Chapter 10, Getting The QSLs).

Contest Strategy For DXers

Competitive contesters want to work as many stations as possible, while also working as many multipliers as are available on each band. DXers usually try to work as many new DXCC entities as possible on each band.

There are two ways to work stations in a contest and your goals will determine how much use to make of each method. You can *search and pounce* (S&P), looking for stations you want to work that are calling CQ, or you can try to find an empty spot on a band and call CQ yourself.

Should You Call CQ?

Competitive stations trying to make large scores spend most of their time calling CQ, occasionally tuning the band in S&P mode. Sitting on one frequency and calling CQ is known as *running*. A proficient operator with a good station can run more than 100 stations per hour. Running stations is fun and can be addictive. As propagation changes, different parts of the world start appearing in your log. You'll work station after station in one country after another. And suddenly you may be thrilled when you're called by a DX entity you've never worked before!

Unless your signal is very strong, though, calling CQ can be fruitless. Rather than successfully running, you're likely to be outgunned by stronger stations also calling CQ. The sheer density of strong signals on the band can bury your weaker signal. S&P may be the only way to make contacts when this happens.

Calling CQ, however, is often the only way to snag some multipliers. Very often the countries or entities most in demand are represented by only one station. That station may not feel up to handling the huge pileup that will result from calling CQ. So he or she tunes through the band calling other stations in S&P mode.

I've often come upon such operators, working their way up the band. I've been able to work them by moving higher in frequency from the station they just worked (it helps to determine if they are tuning higher or lower after each contact!) and calling CQ myself if I can find an open spot.

You'll probably spend most of your time searching and pouncing. Because the bands are usually saturated with signals during a contest, you should practice tuning your transceiver to exactly the same frequency on which the other station is transmitting. This is especially important on CW.

Being *zero beat* greatly improves your chances of being heard and reduces interference with stations on adjacent frequencies.

When To Operate?

Most DX contests run for 48 hours. When solar activity is high and at least two bands are open at all times, planning your operating times can be a real challenge. Here again, your strategy will be different as a DXer, as compared to a contester.

Casual tourists activate many desirable DXCC entities during contests. Some contest DXpeditions, however, are large in scale, with multiple operators on several bands at the same time. To snag the ones staged by single operators you'll want to be operating when propagation favors those locations.

If solar activity is very low and you need a station in Africa, for example, your best bet from the US East Coast will be on 40 or 80/75 meters, meaning you'll have to look for Africa after sunset at your location. If solar activity is high, you'll have more chances to work that station on higher frequencies, which are open during the daylight hours.

It's worth spending time searching the bands during likely times for good propagation to a particular area of the world. You'll probably pick up other new ones in other areas while you're looking. Of course, you can always rely on the DX spotting networks, but where's the fun in that?

There are only a few hours between sunset in the Far East and sunrise in my Southeast US location. If I'm hunting for countries in that region on the low bands (160 to 40 meters), I must be on the air during those hours. I use my propagation prediction program to tell me which band is likely to provide the best chances, but I listen to the other bands too. I scan a band, checking each signal as I tune across it. If I don't find anything new to work, I look for a quiet spot and call a few CQs.

At the time in the late afternoon when I start looking for stations in the Far East, the low bands are open to Europe. During a contest I work the Europeans, but listen carefully for the audible clues that a station calling me is farther away.

When solar activity is high, early mornings on the East Coast bring strong openings to Europe on 20 and 15 meters. But they also bring long path openings to Asia and the Pacific on 10 meters. The Europeans will still be there when the long path closes, so I concentrate on the stations coming from across the Pacific.

When To Change Bands

If your signal is strong enough to run stations, your logging program can suggest when you should change bands. If your rate has been 75 contacts an hour and it drops to 50, you may have worked everything available on that band. When searching and pouncing, I make a couple of passes through the band then change to another band. If you spend some time with your propagation prediction program before the contest, you can plan band changes when propagation favors parts of the world you'd like to work. Here we're assuming you're looking for new DXCC entities, not concentrating on increasing your contact total.

Watching the sunrise-sunset terminator is especially important on the 160, 80/75 and 40 meter bands — most especially on 160 meters. At sunrise or sunset on the other end of a path signal strengths may be briefly enhanced, improving your chances of working a new one on Topband.

Chapter 10

Getting the QSLs

To qualify for DXCC you must submit confirmation that you had two-way communications with authorized Amateur Radio stations (or stations authorized to contact Amateur Radio stations), on authorized Amateur Radio frequencies and using authorized communications modes, with at least 100 DXCC entities or countries.

There are two ways to submit confirmations to ARRL: QSL cards and Logbook of The World (*LoTW*) entries. QSL cards must contain the call sign of the station for which you are claiming credit, the full date and time of the contact, the frequency or band used and the mode (eg, CW, SSB, RTTY). Where there is a possibility of ambiguity (for example, the station's prefix is KP4 but the operation took place from the continental United States, not Puerto Rico), the station's location should also appear on the card.

All contacts submitted for a DXCC award must be made from the same DXCC entity, using only call signs issued to you. Submit QSL cards exactly as you receive them. If the other operator left out a necessary piece of information, you must ask for and then submit a different card. Never modify or correct information on a card. (Remember this when filling out your own cards. If you make a mistake, start fresh with a blank card. Don't cross out and re-enter information.) See Chapter 11, Applying For DXCC And Other Awards, for information on submitting cards and logs for DXCC credit.

Designing A QSL Card

Most commercial QSL card printers supply cards with all necessary blank

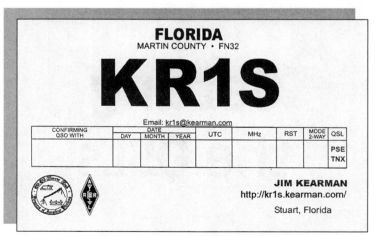

Fig 10-1 — This simple QSL design contains spaces for all required information.

spaces identified. All you have to do is fill in the boxes (make sure you enter the correct date, and always, always use UTC time). Many logging programs can produce printed labels containing all the necessary information, a big help if you've just made several hundred contacts in a contest.

QSL cards come in two varieties, fancy and plain. Fancy cards have photos or colorful graphics, and sometimes the contact information is printed on the back. If you buy fancy cards, make sure your call sign appears on both the front and back of the card. QSL sorters and checkers get crabby when they have to keep flipping cards back and forth. I prefer cards printed on only one side. **Fig 10-1** shows my card, on which I have put the URL of my Web site instead of my street address.

QSL Via the Bureau

Now you know what you need, how do you get those confirmations? Let's start with QSL cards. The easiest and least-expensive way to get QSL cards from DX stations is through your country's QSL bureau. In the United States, the QSL bureau is operated by ARRL. It has two branches, Incoming and Outgoing.

Any radio amateur living in the United States or its territories and possessions may receive DX QSLs via the Incoming ARRL QSL Bureau. To use the Outgoing Bureau, however, you must be an ARRL member. Outgoing QSL cards are sent to ARRL HQ. Incoming QSL cards come to you through individual bureaus in each call area. Use the bureau that corresponds to your call sign prefix. For example, even though I live in the Fourth Call Area, my First Call Area prefix means I use the "W1" Incoming QSL Bureau.

ARRL has posted all you need to know about using the Outgoing QSL Bureau online, at **www.arrl.org/qsl/qslout.html**. This information is subject to change, especially when postage rates go up, so check it frequently. Each Incoming QSL Bureau sets its own requirements. Some will accept stamped envelopes or payment; others will accept only payment, preferring to provide their own standardized envelopes. Find the information for your Incoming QSL Bureau at **www.arrl.org/qsl/qslin.html**.

As postage costs go up, many DX operators prefer to QSL by the bureau system. A unofficial prosign commonly used on CW is "QSLL," meaning, "Let's QSL by the bureau."

QSL Direct

QSL cards sent by the bureau system may take more than a year to arrive. At the time of this writing, more than 70 countries do not offer QSL bureau service. QSLing direct can be expensive and I recommend it only when there is no alternative for getting a QSL card.

If you QSL direct and want a reply by the same method, you must enclose postage or funds of sufficient value. Airmail postage rates in some countries are quite high and not all countries have secure postal agencies. While the American dollar is usually accepted almost anywhere in the world, some countries do not allow their citizens to possess foreign currency. When used for reply postage, the US dollar is often referred to as a *green stamp*. There was a time when one green stamp would cover airmail postage from anywhere, but now many foreign DX stations require two.

Many DXers use International Reply Coupons (IRCs), issued by the Universal Postal Union and available at most post offices. IRC designs have changed over the years and there are thousands of older, obsolete IRCs floating around. I recommend you purchase them only from your post office or from reputable dealers. Some QSL managers (see below) sell excess IRCs; make sure your IRCs are still valid before sending them out. You can get the latest information on IRCs at the Universal Postal Union's Web site, **www.upu.int/**.

I mentioned postal security. Unfortunately, postal workers in some countries supplement their income by stealing currency from incoming mail.

Fig 10-2 — Nothing brightens a DXer's day like receiving an envelope from the QSL bureau!

While you may be proud of your call sign, leave it and the other station's call sign off the envelope. Invest in security envelopes that make it difficult to see the contents when the envelope is held up to the light. Older readers may recall something called carbon paper, which was used in the days of typewriters. I used to cut a piece of carbon paper to fit inside the envelope, carbon (ink) side out, to make the envelope opaque. However you do it, you want to make it hard for thieves to determine that the envelope contains anything of value.

QSL Managers

Many DX stations use QSL managers. A QSL manager receives all that station's logs and incoming QSLs. As cards come in, the manager checks them against the logs and sends reply QSL cards. QSL managers often buy the DX station's QSL cards, and so they charge a fee in addition to reply postage to cover their costs.

DX stations use QSL managers for several reasons. They may be too busy to fill out QSLs or there may be postal security problems in their country. Sometimes their operation is of short duration, so they don't want incoming mail. Very often, the QSL manager call sign is the DX station's home call sign.

Mostly rare or somewhat rare DX stations use QSL managers, but it never hurts to check. Sometimes the DX station will give out the information, saying "QSL via (call sign)." Looking up the DX station's call on your favorite search engine will often return information about his/her QSL manager and about what you need to supply to ensure a reply QSL card.

Some QSL managers will respond to cards sent via the bureau system. In that case, in the upper left corner of your QSL, write "VIA" and the QSL manager's call sign. Note that the Outgoing ARRL Bureau can send cards to QSL managers, so long as those managers are handling cards for DX stations. QSL managers based in the United States can also send reply QSLs via the bureau.

Online QSL Sites

There is at least one non ARRL Web site where you may upload log files and receive electronic QSLs (*eQSLs*). While eQSLs are fun to receive, they are not accepted for ARRL awards, including DXCC. Only electronic confirmations made through ARRL's Logbook Of The World (LoTW) qualify for ARRL awards.

Logbook Of The World

DXing entered the information age with the advent of ARRL's Logbook of

The World (LoTW). Participants upload their logs to ARRL's Web based server, which compares them to find matches based on call signs, date and time, mode and band. Activeoperators upload their logs on a regular basis. If you do the same you may obtain confirmations within minutes of making the contact!

To ensure the integrity of the DXCC program, ARRL has made the LoTW process very secure. Signing up for LoTW may seem intimidating at first, but if you follow the steps it's very easy to do. Step by step instructions follow, but here's a summary.

Confirm that your logging software can output data in a compatible file format, then download and install the small LoTW application program (available for *Windows*, *Mac* and *Linux* operating systems). As part of this process you will request a certificate from the server, which will prompt a postal mail reply from ARRL (**Fig 10-3**). This postcard contains a password. When you receive it you log on to LoTW, enter the password, and you're ready to use the system. Here are some more details about using LoTW.

Getting Started With LoTW

The first step in using LoTW is to make sure your computer logging program is compatible. It must be able to produce files in either ADIF or Cabrillo format. Most current logging programs can; if you're not sure, check

Logbook of the World

Thanks for registering to use Logbook of the World. In order to continue, you must visit the LoTW Web site at **http://www.arrl.org/lotw/** and enter your postcard password. This password is the eight-character group printed above the address on the front of this card. Enter it exactly as printed.

If you cannot read the password please call 860-594-0206 and ask for a new postcard to be sent. The postcard must be sent to the address listed with the FCC for your call sign.

If you enter your password, but it is not accepted, please verify that you have correctly read and entered the password. If it still does not work, you may send an email to **lotw-help@arrl.org** alerting us, or call 860-594-0206.

Having successfully entered your password, you will receive your certificate as an email attachment within three business days.

Fig 10-3 — An example of the postcard that US LoTW applicants will receive from ARRL. The password used to request an e-mailed certificate file is printed on the front. Applicants from outside the USA must submit documentation by postal mail and then they can receive their certificate files by e-mail.

the program's Help files or contact the software vendor.

Radio Amateurs in the United States should make sure their mailing address is the same as the address in the FCC database. Use the online call sign database at **www.arrl.org/** to check and make corrections as necessary, *before* beginning the LoTW sign up process. ARRL will only mail your confirmation to the address in the FCC database.

Your next step is to go online and download two small programs, *TQSL* and *TQSL Cert*. You'll find them at **www.arrl.org/lotw/**. Separate versions are provided for *Windows* (95 and later), *MacOS X* and *Linux*. The LoTW software is free. The *Windows* download is about 2.2 MB. It is a self extracting executable, so double-click on the icon and follow the prompts. *Linux* users must download two RPMs; *MacOS X* users download a single file.

Of the two programs in the download(s), the one you'll need first is *TQSL Cert*. That program creates a file that you will e-mail to ARRL. The e-mail address is **lotw-logs@arrl.org**. If you live in the United States, when ARRL receives your e-mail ARRL will send you a postcard by snail mail with a password, as explained earlier. You will be asked for an e-mail address, to which ARRL will send a confirmation and also the certificate file itself. Make sure you enter a valid e-mail address, and that your e-mail account spam filters will not block replies or e-mails containing attachments.

Because ARRL cannot electronically authenticate the identity of non-USA radio amateurs, hams outside the USA must request LoTW certification by postal mail. The requirements for certification are online at **p1k.arrl.org/lotw/docreq**, and also in the sidebar below.

When you receive your postcard from ARRL (non-USA stations will not get a postcard; their certificates will be sent directly by e-mail, so they should skip this step), go to **www.arrl.org/lotw/**. Under "LoTW Links," click on "Enter a password from a post card" and enter the call sign for which you requested a certificate. Next enter the password printed on the front of the postcard. In short order you will receive an e-mail with your certificate file attached. This e-mail also contains the password you will use to log on to the LoTW User's site to upload and review your logs.

Your certificate file is important, so make a backup copy on a floppy or CD-R and keep it in a safe location. If you change computers or upgrade the *TQSL* application, you will need that certificate file. If you lose your certificate file, send an e-mail to **lotw-help@arrl.org** and ask for assistance.

Now you're finally ready to upload log files. To do that, you'll need a log file in ADIF or Cabrillo format. (You can use *TQSL* to create ADIF files from raw data but the process is laborious; use your logging program instead.) My logging program has a function to "Export ADIF Format," and lets me select either the entire log or selected entries for export.

Non-USA Requirements for LoTW Certificates

• Every **non-USA operator** requesting an initial, unsigned certificate for Logbook of the World must send a copy of his/her Amateur Radio operating authorization, in addition to a copy of one other government-issued document indicating his/her identity. Such an additional document might be a driver's license, or the first page of a passport. (Identity documents will be destroyed at ARRL HQ after certificates are successfully issued.) *These documents must be sent via postal mail (to the ARRL HQ address below), not e-mail.* Using the postal mail for documenting an operator's primary call sign helps to protect your call sign as well as helping LoTW to authenticate its users. ARRL recognizes that non-USA operators may object to sending personal information through the mails, and is working toward local confirmation instead.

• **Documentation should be mailed to***:*

Logbook Administration
ARRL
225 Main Street
Newington, CT 06111
USA

• Each **non-USA operator**, who already has a certificate, may request additional certificates for other callsigns, such as old calls or DXpedition calls, etc. No additional *identity documentation is required, but for each additional certificate, the applicant must submit proof of operating authorization unless:*

1. The operation takes place under CEPT or IARP conventions, or

2. The operation takes place under an agreement between two or more countries whereby one DOES NOT need to have an additional, written license or operating authority.

• Reciprocal operating authority between two countries, outside of CEPT or other agreements, does not eliminate *LoTW*'s documentation requirement. In cases where old documentation is no longer available, you may contact ARRL for guidance. Identity documentation is not required when the additional request is being signed by an existing certificate.

• If you have already obtained a certificate, and you are required to send documentation (a copy of a license) to *LoTW* for additional callsigns, it is OK to scan this license and e-mail it to LoTW-help@arrl.org. *Remember that you can only e-mail license images for additional certificate requests — not for your first certificate.*

You won't upload your complete log every time, only contacts made since your last upload. You can upload contacts as you make them (confirming them almost before you've signed off with the other station on the air!), once a day or whenever convenient. (Some logging programs will work with *TQSL* to directly upload and download files to and from LoTW.)

When you've exported the file, run *TQSL*. From the File menu, select "Sign Existing ADIF or Cabrillo file." You'll be asked to select a station location. Most of us have only one station location, but it is possible to certify several locations under one certificate and have separate LoTW logs for each one.

Navigate to the directory where you saved the exported log file and select the file. *TQSL* will ask for a filename, because it is going to make a copy of your log in LoTW format, digitally signed from your certificate. You'll be asked to confirm the station location for which you are creating the LoTW file. Again, since most users have only one station location there is nothing to confirm. If you have certified multiple station locations this is your last chance to make sure you select the right one.

The next option is to select starting and ending dates from the log file. If you didn't select the dates of contacts to be uploaded in your logging program, you can do that now. For example, if you exported your entire log, but only want to upload contacts made since your last upload (recommended), enter the dates. If you are submitting your log for the first time, or have exported only contacts made since your last upload, leave the date fields blank.

Depending on the size of your log it may take several minutes for *TQSL* to prepare the LoTW file. If there are date errors in your log (you accidentally entered the year as 2016 instead of 2006, for example), they are listed in the window as the log file is created. If you find errors you should correct them *in your logging program*, export the log to ADIF format again and run *TQSL* again.

When you have created an error-free log, you're ready to e-mail or upload it to ARRL. I upload my files by selecting the Upload link at **p1k.arrl.org/lotwuser/**. You can also e-mail your signed log file to **lotw-logs@arrl.org**, but I recommend using the Web server. If you are uploading a log for an ARRL contest, you must submit the log twice: Once in Cabrillo format (to the specific contest e-mail address) for the contest entry itself, and once again in signed LoTW format for LoTW.

Now that you've uploaded your logs, you're ready to see which contacts are already confirmed. It takes from several minutes to a couple of hours for the ARRL server to integrate your logs and verify contacts. The delay is longer after major contests. You access your logs at the *secure* page **https://p1k.arrl.org/lotw**. (You must type "https," not "http.")

Click on "Log onto the LoTW User's Page" and log in using your call sign and the password ARRL e-mailed back to you. Now you are looking at the

LoTW User's Home Page. This page contains much useful information, which you should read before continuing. Then click on the "Your QSOs" link.

You'll see a query form you can use to list all contacts you have uploaded, or only contacts between certain dates, on a particular band or mode, or listed by DXCC entity. As you probably want to know which entities are confirmed in LoTW, check the "Show Confirmed QSOs only" box. If you want to download the report generated from the form, click the "Download Report" link. Here you can select a start date for the report and choose to have the report indicate which QSOs are confirmed. You'll be prompted to save the file to your computer. This file is in ADIF format, suitable for importing into your log program.

LoTW saves DXers a great deal of time and money. Many DXers have achieved DXCC on several bands using only LoTW, but you'll still need some paper QSL cards. If you frequently update and check your LoTW records you'll know which stations will require paper QSLs.

Linking LoTW To Your DXCC Account

If you have previously qualified for DXCC from the DXCC entity from which you are using LoTW, you can link to your DXCC account. DXCC applications for new or endorsed awards made since 1991 are computerized. When you log into the LoTW User's Home Page, select the Awards link. Most of us have only one LoTW account. If not, select the one you want to link to a particular existing DXCC record, then click Select DXCC Award Account. From there you can Link To your computerized records, merging them with your LoTW records.

Record Keeping

Now is the time to develop a system for keeping track of DXCC entities worked and confirmed. Computer logging software is the easiest method. If you prefer paper records you'll want to track contacts by band and operating mode, as well as your mixed-mode, all-band totals. That means keeping at least three lists. Logging software saves many headaches but you must maintain backup copies of all your logs. LoTW will provide a safe backup for your contact logs and contacts confirmed by LoTW. Contacts confirmed by paper QSL cards will not be documented online until you submit the cards for DXCC credit. You can track paper QSLs in your logging program, but a backup copy of that file is still essential.

Your first DXCC application will probably be for the Mixed Award (Chapter 1), but eventually you will want to apply for single mode or single band awards. Whether you keep manual records or use a logging program, set aside the time to update and back up your records!

Chapter 11

Applying for DXCC and Other Awards

The day you've waited for has arrived! Between LoTW and paper QSLs you now have confirmations from 100 DXCC entities. Now it's time to apply for that beautiful certificate. While the DXCC Rules are included in this book, changes occur frequently. Refer to **www.arrl.org/awards/dxcc/ rules.html** for the latest rules.

DXCC Application Procedure

There are two ways to apply for DXCC. You can submit your cards directly to ARRL HQ, or have them checked by an authorized DXCC Card Checker at a convention or hamfest. (Applications using *only* LoTW-confirmed contacts may be submitted online.) Here's what you'll need to submit your first application for DXCC.

- A DXCC Award Application. You can download the form at **www. arrl.org/awards/dxcc/**. You must use this form for all DXCC applications, whether for a new award or to endorse an existing award.

- The QSL cards you will submit, properly sorted. This is an important part of the application process. Your cards must be arranged first by band, then by mode. Cards with more than one QSO should be placed at the end. Cards must not be sorted alpha-numerically.

- A DXCC Record Sheet, which you can download from **www.arrl. org/awards/dxcc/dxccrec.pdf**. While this form is not required if you send your cards directly to ARRL HQ for checking, it's helpful when sorting your cards. If you plan to submit your cards to an authorized DXCC Card Checker, you must provide the filled-out form with your cards.

- A form of payment. ARRL charges a small fee for processing an application. If you are submitting cards to ARRL HQ you must pay for return postage. Also, except for your first-ever DXCC application, there is a nominal additional charge for the DXCC certificate. ARRL accepts money orders or personal checks payable in US dollars, US currency (not recommended if you are mailing the cards), or major credit cards.

- US applicants must be ARRL members at the time of each application for a new DXCC award or endorsement of existing DXCC awards. You'll need to know your membership expiration date, but your membership number is not required.

Filling Out The Application

The first block in the upper-left corner of the application is where you tell the DXCC Desk which award you want. Assuming this is your first DXCC application, you will probably check the box corresponding to New Award and MIX (Mixed Mode). The mode options are Phone, CW, RTTY and Satellite. Or you may be applying for a single-band award. It's not likely your first application will be for Five-Band DXCC (5BDXCC), but if it is, check that box.

Your first-ever DXCC application includes a DXCC certificate at no additional charge. Subsequent applications for which you want a separate certificate require payment of an additional fee. You may apply for these awards and not request a certificate, but if that's what you intend, do not check the New boxes. Checking the New box will cause a new certificate to be prepared, for which you will be charged.

Go down the page and enter the number of QSL cards you are submitting with this application, and the number of QSOs for which you are claiming credit. If you worked a station more than once on different bands and modes, all QSOs you had with that station may be listed on one QSL card. For the Mixed Award you can only claim credit for a single entity once, even if you worked it on more than one band or mode. If you are applying for multiple single-band or single-mode awards or 5BDXCC, you may refer to multiple contacts on a single card. All cards covering more than one QSO should be placed at the bottom of the stack of cards you are submitting, even if you are only claiming credit for one QSO.

Even if you are only claiming credit for one QSO on a multi-QSO card, you should tell DXCC which one you are claiming. Do this with a Post-It® note or a slip of paper taped to the card. *Never write on a QSL card.* Unless you are sending sufficient confirmations for additional awards (and have enclosed the correct fees), you cannot claim credit "in advance" for other QSOs on the same card. If you are applying for two single-band awards, for example, each award has a separate application fee, and your application must be accompanied by all the confirmations needed for both awards.

Continuing in the upper-right corner of the application, enter your full name, the call sign for which you are applying for DXCC (this is the call sign that will be printed on the certificate), and any other calls you held in the past that may appear on QSLs you are submitting. Note that you can only claim credit for an award for contacts you made from within a single DXCC entity.

Now enter the mailing address where you want your cards returned, and where your certificate (if you request one) will be sent. If you don't want your cards returned, write "Do not return cards" here, but do enter a mailing address if you want a certificate.

You don't have to give a phone number or e-mail address, but if ARRL has questions about your application, being able to contact you quickly will expedite the application process. Now you have to do some math. For your first application, the cost for US ARRL members is $12, which includes the certificate, but does not include the cost of returning the cards. If you are paying by check or money order you'll have to estimate the cost. You must specify how you want your cards shipped, or ARRL will send them via Registered Mail. This can be expensive for overseas applicants. The cost for non-US ARRL members is also $12, and $22 for non-members. US applicants must be ARRL members, non-US applicants do not. ARRL cannot bill you for application fees, so be sure to remit the correct amount when you apply.

Sign and date the application. And don't forget to include your ARRL Membership expiration date (Life Members enter "Life Member") if applicable.

If you're submitting the cards to ARRL, review the checklist at the beginning of this section to make sure you haven't forgotten anything. Box up your cards and application, make sure your form of payment is enclosed, and mail or ship it to:

ARRL DXCC
225 Main St
Newington, CT 06111
USA

Having Cards Checked At Hamfests

Thanks to dozens of volunteers around the world, you may not have to send your cards to ARRL HQ for checking. Certain cards are not eligible, however, to be checked by DXCC Card Checkers. These are 160-meter cards and cards for deleted entities. DXCC Card Checkers are themselves DXCC members. You'll find one or more at almost every hamfest and Amateur Radio convention in the United States and Canada. There are DXCC Card Checkers in almost two dozen other countries as well. DXCC Card Checkers are often members of local DX and contest clubs. A list of current DXCC Card Checkers is online at **www.arrl.org/awards/dxcc/checkers.html**. When you're ready to submit your application, contact a Card Checker near you to see when you can get together. You may be able to meet at a club meeting as well, and that's a good chance to meet other DXers.

Make sure your application is completely filled in before meeting the Card Checker. You'll also need a filled-in DXCC Record Sheet, proof of membership if you live in the US, and a means of paying the required DXCC fees. A check or money order payable in US currency, or a credit card is probably preferred by most Card Checkers, so they don't have to handle cash. When your cards are checked they'll be returned to you, and your application and payment will be sent to ARRL HQ within two days. Make sure to give your Card Checker a big "Thanks" for volunteering his or her time!

What You'll Get Back From ARRL

Application processing normally takes between 3-5 weeks. You can check the status of your application by going to the DXCC List of Applications Received page at **www.arrl.org/awards/dxcc/appstatus.html**. If this is your first application, you'll be receiving a certificate, which will come in a mailing tube, so be on the lookout for that. If your application shows up on the status page and you provided a valid e-mail address, phone number, or both, no news is good news. The DXCC Desk works as quickly as possible to process applications, so be patient!

In addition to your certificate and your cards (if you didn't specify that you didn't want them returned), you'll receive a DXCC Credit Slip. It shows all entities for which you have received credit, a list of any submitted cards for which you did not receive credit and why they were rejected. Another section covers fees for which you were charged.

The second and largest part of your report is your record of countries worked, with the X's indicating credits received. When you receive this report it is very important that you check it against your personal records for accuracy. Before reporting problems you should also verify the results against your personal log.

ARRL maintains a Frequently Asked Questions (FAQ) page for DXCC applicants, at **www.arrl.org/awards/dxcc/faq/**.

Other DX Awards

DXCC is the most-prestigious DX award in the amateur world, but there are other challenges you'll want to take on as you progress.

ARRL DXCC Challenge And The DeSoto Cup

The *DXCC Challenge* recognizes your combined total for all DX bands from 160 to 6 meters. A minimum of 1000 band-entities is required. All contacts made and confirmed since November 15, 1945, may be counted for credit. This includes only current entities. Deleted entities do not count towards this award. Once you reach the 1000-entity level, you are entered into the Challenge listing automatically. An application to request the DXCC Challenge listing is not required. This award is endorsable in levels of 500. There is no certificate for this award. A special plaque has been designed for the DXCC Challenge. Confirmations that have already been submitted to DXCC may be claimed as well, but only current entities count for this award. The "Chal" column on your DXCC Credit slip indicates your current Challenge total.

You can find a list of those DXers who already have more than 1000 Challenge credits on the DXCC Web site. When your total reaches 1000 band-entities, a letter of congratulations will accompany your submission results.

The unique Challenge plaque is endorsable in increments of 500 and comes with the first endorsement "medallion." A distinctive lapel pin indicating your Challenge level is also available. Check the DXCC Web site for additional information and a printable order form. More information about the DXCC Challenge is online at **www.arrl.org/news/stories/2001/05/08/3/**.

Worked All Continents (WAC)

WAC is usually the first award beginning DXers apply for. WAC is even older than DXCC, tracing its roots back to the founding of the International Amateur Radio Union (IARU) in 1925. To qualify for WAC you must confirm two-way communications with Amateur Radio stations on each of the six continents: Africa, Asia, Europe, North America, Oceania and South America. Application forms are available online at **www.iaru.org/wac/wac.pdf**. US Amateurs and those in countries without a national Amateur Radio society may apply through ARRL HQ (ARRL is the Headquarters Society of IARU). Amateurs in other countries with national Amateur Radio organization should apply through them. Applicants must be members of their national amateur radio societies affiliated with IARU, unless no national society exists in their countries.

In addition to the basic WAC award, separate awards are available for 5- and 6-Band WAC, and QRP WAC. See the IARU WAC home page, **www.iaru.org/wac/**, for more information.

Worked All Zones (WAZ)

WAZ is another prestigious DX award, sponsored by *CQ Magazine*. The 40 *CQ* Zones are not the same as ITU (International Telecommunications Union) Zones. Most WAZ zones are easy to work, but some are devilishly difficult. Separate awards are available for different modes and bands, much like the DXCC Award. Complete rules and application forms are available from **www.cq-amateur-radio.com/wazrules.html**.

National Society And Club Awards

Many other national Amateur Radio societies and Amateur Radio clubs offer attractive awards. In most cases these awards require contacts with a certain minimum number of society or club members. Good places to learn more about the many awards available are **www.ac6v.com/hamawards.htm** and **www.dxawards.com/links.htm**, and the monthly Awards column in *CQ* (**www.cq-amateur-radio.com/**).

Appendix

DXCC Rules

DXCC Rules are subject to change. While these rules were current at the time of publication, refer to **www.arrl.org/awards/dxcc/rules.html** for the latest version.

SECTION I. BASIC RULES

1. The DX Century Club Award, with certificate and lapel pin is available to Amateur Radio operators throughout the world (see #15 below for the DXCC Award Fee Schedule). ARRL membership is required for DXCC applicants in the US, its possessions, and Puerto Rico. ARRL membership is not required for foreign applicants. All DXCCs are endorsable (see Rule 5). There are 18 separate DXCC awards available, plus the DXCC Honor Roll:

a) Mixed (general type): Contacts may be made using any mode since November 15, 1945.

b) Phone: Contacts must be made using radiotelephone since November 15, 1945. Confirmations for cross-mode contacts for this award must be dated September 30, 1981, or earlier.

c) CW: Contacts must be made using CW since January 1, 1975. Confirmations for cross-mode contacts for this award must be dated September 30, 1981, or earlier.

d) RTTY: Contacts must be made using radioteletype since November 15, 1945. (Baudot, ASCII, AMTOR and packet count as RTTY.) Confirmations for cross-mode contacts for this award must be dated September 30, 1981, or earlier.

e) 160 Meter: Contacts must be made on 160 meters since November 15, 1945.

f) 80 Meter: Contacts must be made on 80 meters since November 15, 1945.

g) 40 Meter: Contacts must be made on 40 meters since November 15, 1945.

h) 30 Meter: Contacts must be made on 30 meters since November 15, 1945.

i) 20 Meter: Contacts must be made on 20 meters since November 15, 1945.

j) 17 Meter: Contacts must be made on 17 meters since November 15, 1945.

k) 15 Meter: Contacts must be made on 15 meters since November 15, 1945.

l) 12 Meter: Contacts must be made on 12 meters since November 15, 1945.

m) 10 Meter: Contacts must be made on 10 meters since November 15, 1945.

n) 6 Meter: Contacts must be made on 6 meters since November 15, 1945.

o) 2 Meter: Contacts must be made on 2 meters since November 15, 1945.

p) Satellite: Contacts must be made using satellites since March 1, 1965. Confirmations must indicate satellite QSO.

q) Five-Band DXCC (5BDXCC): The 5BDXCC certificate is available for working and confirming 100 current DXCC entities (deleted entities don't count for this award) on each of the following five bands: 80, 40, 20, 15, and 10 Meters. Contacts are valid from November 15, 1945.

The 5BDXCC is endorsable for these additional bands: 160, 30, 17, 12, 6, and 2 Meters. 5BDXCC qualifiers are eligible for an individually engraved plaque (at a charge of $40.00 US plus shipping).

r) The DXCC Challenge Award is given for working and confirming at least 1,000 DXCC band-Entities on any Amateur bands, 1.8 through 54 MHz (except 60 meters). This award is in the form of a plaque. Certificates are not available for this award. Plaques can be endorsed in increments of 500. Entities for each band are totaled to give the Challenge standing. Deleted entities do not count for this award. All contacts must be made after November 15, 1945. The DXCC Challenge plaque is available for $79.00 (plus postage). QSOs for the 160, 80, 40, 30, 20, 17, 15, 12, 10 and 6 meter bands qualify for this award. Bands with less than 100 contacts are acceptable for credit for this award.

The DeSoto Cup

s) The DeSoto Cup is presented to the DXCC Challenge leader as of the 31st of December each year. The DeSoto Cup is named for Clinton B. DeSoto, whose definitive article in October 1935 QST forms the basis of the DXCC award. Only one cup will be awarded to any single individual. A medal will be presented to a repeat winner in subsequent years. Medals will also be awarded to the second and third place winners each year.

t) Honor Roll: Attaining the DXCC Honor Roll represents the pinnacle of DX achievement:

i) Mixed: To qualify, you must have a total confirmed entity count that places you among the numerical top ten DXCC entities total on the current DXCC List (example: if there are 337 current DXCC entities, you must have at least 328 entities confirmed).

 ii) Phone -- same as Mixed.

 iii) CW -- same as Mixed.

 iv) RTTY -- same as Mixed.

To establish the number of DXCC entity credits needed to qualify for the Honor Roll, the maximum possible number of current entities available for credit is published monthly on the ARRL DXCC Web Page. First-time Honor Roll members are recognized monthly on the DXCC Web Page. Complete Honor Roll standings are published annually in QST, usually in the August issue. See DXCC notes in QST for specific information on qualifying for this Honor Roll standings list. Once recognized on this list or in a subsequent monthly update of new members, you retain your Honor Roll standing until the next standings list is published. In addition, Honor Roll members who have been listed in the previous Honor Roll Listings or have gained Honor Roll status in a subsequent monthly listing are recognized in the DXCC Annual List. Honor Roll qualifiers receive an Honor Roll endorsement sticker for their DXCC certificate and are eligible for an Honor Roll lapel pin ($6) and an Honor Roll plaque ($40 plus shipping). Write the DXCC Desk for details or check out the Century Club Item Order Form at /awards/dxcc/

v) #1 Honor Roll: To qualify for a Mixed, Phone, CW or RTTY Number One plaque, you must have worked every entity on the current DXCC List. Write the DXCC Desk for details. #1 Honor Roll qualifiers receive a #1 Honor Roll endorsement sticker for their DXCC certificate and are eligible for a #1 Honor Roll lapel pin ($6) and a #1 Honor Roll plaque ($55 plus shipping).

 2. Written Proof: Except in cases where the rules of Section IV apply, proof of two-way communication (contacts) must be submitted directly to ARRL Headquarters for all DXCC credits claimed. Photocopies and electronically transmitted confirmations (including, but not limited to fax, telex and telegram) are not currently acceptable for DXCC purposes.

Exception: Confirmations created and delivered by ARRL's Logbook of the World system are acceptable for DXCC credit.

The use of a current official DXCC application form or an approved facsimile (for example, exactly reproduced by a computer program) is required. Such forms must include provision for listing callsign, date, band, mode, and DXCC entity name. Complete application materials are available from ARRL Headquarters. Confirmations for a total of 100 or more different DXCC credits must be included with your first application.

Cards contained in the original received envelopes or in albums will be returned at applicant's expense without processing.

3. The ARRL DXCC List is based on the DXCC List Criteria.

4. Confirmation data for two-way communications must include the call signs of both stations, the Entity name as shown in the DXCC List, mode, and date, time and band. Except as permitted in Rule 1, cross-mode contacts are not permitted for DXCC credits. Confirmations not containing all required information may be rejected.

5. Endorsement stickers for affixing to certificates or pins will be awarded as additional DXCC credits are granted. For the Mixed, Phone, CW, RTTY, 40, 30, 20, 17, 15, 12 and 10-Meter DXCC, stickers are provided in exact multiples of 50, i.e. 150, 200, etc. between 100 and 250 DXCC credits, in multiples of 25 between 250 and 300, and in multiples of 5 above 300 DXCC credits.

For 160-Meter, 80-Meter, 6-Meter, 2-Meter and Satellite DXCC, the stickers are issued in exact multiples of 25 starting at 100 and multiples of 10 above 200, and in multiples of 5 between 250 and 300. Confirmations for DXCC credit may be submitted in any increment, but stickers and listings are provided only after a new level has been attained.

6. All contacts must be made with amateur stations working in the authorized amateur bands or with other stations licensed or authorized to work amateurs. Contacts made through "repeater" devices or any other power relay methods (other than satellites for Satellite DXCC) are invalid for DXCC credit.

7. Any Amateur Radio operation should take place only with the complete approval and understanding of appropriate administration officials.

In countries where amateurs are licensed in the normal manner, credit may be claimed only for stations using regular government-assigned call signs or portable call signs where reciprocal agreements exist or the host government has so authorized portable operation. Without documentation supporting the operation of an amateur station, credit will not be allowed for contacts with such stations in any country that has temporarily or permanently closed down Amateur Radio operations by special government edict or policy where amateur licenses were formerly issued in the normal manner. In any case, credit will be given for contacts where adequate evidence of authorization by appropriate authorities exists, notwithstanding any such previous or subsequent edict or policy.

8. All stations contacted must be "land stations." Contacts with ships and boats, anchored or underway, and airborne aircraft, cannot be counted. For the purposes of this award, remote control operating points must also be land based. Exception: Permanently docked exhibition ships, such as the Queen Mary and other historic ships will be considered land based.

9. All stations must be contacted from the same DXCC Entity. The location of any station shall be defined as the location of the transmitter. For the purposes of this award, remote operating points must be located within the same DXCC Entity as the transmitter and receiver.

10. All contacts must be made using callsigns issued to the same station licensee. Contacts made by an operator other than the licensee must be made from a station owned and usually operated by the licensee and must be made in accordance with the regulations governing the license grant. Contacts may be made from other stations provided they are personally made by the licensee. The intent of this rule is to prohibit credit for contacts made for you by another operator from another location. You may combine confirmations from several callsigns held for credit to one DXCC award, as long as the provisions of Rule 9 are met. Contacts made from club stations using a club callsign may not be used for credit to an individual's DXCC.

11. All confirmations must be submitted exactly as received by the applicant. The submission of altered, forged, or otherwise invalid confirmations for DXCC credit may result in disqualification of the applicant and forfeiture of any right to DXCC membership. Determinations by the ARRL Awards Committee concerning submissions or disqualification shall be final. The ARRL Awards Committee shall also determine the future eligibility of any DXCC applicant who has ever been barred from DXCC.

12. Conduct: Exemplary conduct is expected of all amateur radio operators participating in the DXCC program. Evidence of intentionally disruptive operating practices or inappropriate ethical conduct in any aspect of DXCC participation may lead to disqualification from all participation in the program by action of the ARRL Awards Committee.

Actions that may lead to disqualification include, but are not limited to:

a) The submission of forged or altered confirmations.

b) The presentation of forged or altered documents in support of an operation.

c) Participation in activities that create an unfavorable impression of amateur radio with government authorities. Such activities include malicious attempts to cause disruption or disaccreditation of an operation.

d) Blatant inequities in confirmation (QSL) procedures. Continued refusal to issue QSLs under certain circumstances may lead to disqualification. Complaints relating to monetary issues involved in QSLing will generally not be considered, however.

13. Each DXCC applicant, by applying, or submitting documentation, stipulates to:

a) Having observed all pertinent governmental regulations for Amateur Radio in the country or countries concerned.

b) Having observed all DXCC rules.

c) Being bound by the DXCC rules.

d) Being bound by the decisions of the ARRL Awards Committee.

14. All DXCC applications (for both new awards and endorsements) must include sufficient funds to cover the cost of returning all confirmations (QSL cards) via the method selected. Funds must be in US dollars using US currency, check or money order made payable to the ARRL, or credit card number with expiration date. Address all correspondence and inquiries relating to DXCC awards and all applications to: ARRL Headquarters, DXCC Desk, 225 Main St., Newington, CT 06111, USA, or send an e-mail to **dxcc@arrl.org**.

15. Fees. All amateurs applying for their very first DXCC Award will be charged a one-time registration fee. For ARRL Members this fee is $12. For foreign non-ARRL members this fee is $22. All first time applicants will receive one free DXCC certificate and lapel pin. Applicants must provide adequate return postage for QSL cards.

a) A $12.00 fee will be charged for each additional DXCC certificate issued, whether new or replacement.

b) Applications may be presented in person at ARRL HQ, and at certain ARRL conventions. When presented at conventions in this manner, such applications shall be limited to 120 QSOs maximum, and an additional $7.00 fee will apply. Applications processed "while-you-wait" at ARRL HQ will be accessed an additional $9.00 fee.

c) Each ARRL member will be charged $12.00 plus postage for the first submission of the year, up to 120 QSOs.

d) Foreign non-ARRL members will be charged $22.00 plus postage for their first submission in a calendar year. Fees in 15. a), 15. b), and 15. d), and 15. f) also apply.

e) A $0.15 fee will be charged for each QSO credited beyond the limits described in 15. b), 15. c), and 15. f).

f) DXCC participants who wish to submit more than once in a calendar year will be charged an additional $10 fee for each second or subsequent application. The fees for a second or subsequent submission in a calendar year are $22 for ARRL members and $32 for foreign non-ARRL members. For a limited time, no Additional $10 fee is charged for second and subsequent applications if LoTW credits are included. The $12 application fee still applies. Return postage must be provided by applicant. Charges from 15. a), 15. b) and 15. d) apply.

16. The ARRL DX Advisory Committee (DXAC) requests your comments and suggestions for improving DXCC. Address correspondence, including petitions for new listing consideration, to ARRL Headquarters, DXAC, 225 Main St., Newington, CT 06111, USA. The DXAC may be contacted by e-mail via the DXCC Desk at **dxcc@arrl.org**. Correspondence on routine DXCC matters should be addressed to the DXCC Desk, or by e-mail to **dxcc@arrl.org**.

SECTION II. DXCC LIST CRITERIA

Introduction:

The ARRL DXCC List is the result of progressive changes in DXing since 1945. Each Entity on the DXCC List contains some definable political or geographical distinctiveness. While the general policy for qualifying Entities for the DXCC List has remained the same, there has been gradual evolution in the specific details of criteria which are used to test Entities for their qualifications. The full DXCC List does not conform completely with current criteria, for some of the listings were recognized from pre-WWII or were accredited with earlier versions of the criteria. In order to maintain continuity with the past, as well as to maintain a robust DXCC List, all Entities on the List at the time the 1998 revision became effective were retained.

Definitions:

Certain terms occur frequently in the DXCC criteria and are listed here. Not all of the definitions given are used directly in the criteria, but are listed in anticipation of their future use.

Entity: A listing on the DXCC List; a counter for DXCC awards. Previously denoted a DXCC "Country."

Event: An historical occurrence, such as date of admission to UN or ITU that may be used in determining listing status.

Event Date: The date an Event occurs. This is the Start Date of all Event Entities.

Event Entity: An Entity created as the result of the occurrence of an Event.

Discovery Entity: An Entity "Discovered" after the listing is complete. This applies only to Geographic Entities, and may occur after a future rule change, or after an Event has changed its status.

Discovery Date: Date of the rule change or Event which prompts addition of the Entity. This is the Start Date for a Discovery Entity.

Original Listing: An Entity which was on the DXCC List at the time of inception.

Start Date: The date after which confirmed two-way contact credits may be counted for DXCC awards.

Add Date: The date when the Entity will be added to the List, and cards will be accepted. This date is for administrative purposes only, and will occur after the Start Date.

Island: A naturally formed area of land surrounded by water, the surface of which is above water at high tide. For the purposes of this award, it must consist of connected land, of which at least two surface points must be separated from each other by not less than 100 meters measured in a straight line from point to point. All of the connected land must be above the high tide mark, as demonstrated on a chart of sufficient scale. For the purposes of this award, any island, reef, or rocks of less than this size shall not be considered in the application of the water separation criteria described in Part 2 of the criteria.

Criteria:

Additions to the DXCC List may be made from time to time as world conditions dictate. Entities may also be removed from the List as a result of political or geographic change. Entities removed from the List may be returned to the List in the future, should they requalify under this criteria. However, an Entity requalified does so as a totally new Entity, not as a reinstated old one.

For inclusion in the DXCC List, conditions as set out below must be met. Listing is not contingent upon whether operation has occurred or will occur, but only upon the qualifications of the Entity.

There are five parts to the criteria, as follows:

1. Political Entities
2. Geographical Entities
3. Special Areas
4. Ineligible Areas
5. Removal Criteria

1. Political Entities:

Political Entities are those areas which are separated by reason of government or political division. They generally contain an indigenous population which is not predominantly composed of military or scientific personnel.

An Entity will be added to the DXCC List as a Political Entity if it meets one or more of the following criteria:

a) The entity is a UN Member State.

b) The entity has been assigned a callsign prefix bloc by the ITU. (The exceptions to this rule are international organizations, such as the UN and ICAO. These Entities are classified under Special Areas, 3.a); and Ineligible Areas, 4.b).) A provisional prefix bloc assignment may be made by the Secretary General of ITU. Should such provisional assignments not be ratified later by the full ITU, the Entity will be removed from the DXCC List.

c) The Entity contains a permanent population, is administered by a local government, and is located at least 800 km from its parent. To satisfy the "permanent population" and "administered by a local government" criteria of this sub-section, an Entity must be listed on either (a) the U.S. Department of State's list of "Dependencies and Areas of Special Sovereignty" as having a local "Administrative Center," or (b) the United Nationslist of "Non-Self-Governing Territories."

New Entities satisfying one or more of the conditions above will be added to the DXCC List by administrative action as of their "Event Date."

Entities qualifying under this section will be referred to as the "Parent" when considering separation under the section "Geographical Separation." Only Entities in this group will be acceptable as a Parent for separation purposes.

2. Geographic Separation Entity:

A Geographic Separation Entity may result when a single Political Entity is physically separated into two or more parts. The part of such a Political Entity that contains the capital city is considered the Parent for tests under these criteria. One or more of the remaining parts resulting from the separation may then qualify for separate status as a DXCC Entity if they satisfy paragraph a) or b) of the Geographic Separation Criteria, as follows.

a) Land Areas:

A new Entity results when part of a DXCC Entity is separated from its Parent by 100 kilometers or more of land of another DXCC Entity. Inland waters may be included in the measurement. The test for separation into two areas requires that a line drawn along a great circle in any direction from any part of the proposed Entity must not touch the Parent before crossing 100 kilometers of the intervening DXCC Entity.

b) Island Areas (Separation by Water):

A new Entity results in the case of an island under any of the following conditions:

i) The island is separated from its Parent, and any other islands that make up the DXCC Entity that contains the Parent, by 350 kilometers or more. Measurement of islands in a group begins with measurement from the island containing the capital city. Only one Entity of this type may be attached to any Parent.

ii) The island is separated from its Parent by 350 kilometers or more, and from any other island attached to that Parent in the same or a different island group by 800 kilometers or more.

iii) The island is separated from its Parent by intervening land or islands that are part of another DXCC Entity, such that a line drawn along a great circle in any direction, from any part of the island, does not touch the Parent before touching the intervening DXCC Entity. There is no minimum separation distance for the first island Entity created under this rule. Additional island Entities may be created under this rule, provided that they are similarly separated from the Parent by a different DXCC Entity and separated from any other islands associated with the Parent by at least 800 km.

3. Special Areas:

The Special Areas listed here may not be divided into additional Entities under the DXCC Rules. None of these constitute a Parent Entity, and none creates a precedent for the addition of similar or additional Entities.

a) The International Telecommunications Union in Geneva (4U1ITU) shall, because of its significance to world telecommunications, be considered as a Special Entity. No additional UN locations will be considered under this ruling.

b) The Antarctic Treaty, signed on 1 December 1959 and entered into force on 23 June 1961, establishes the legal framework for the management of Antarctica. The treaty covers, as stated in Article 6, all land and ice shelves below 60 degrees South. This area is known as the Antarctic Treaty Zone. Article 4 establishes that parties to the treaty will not recognize, dispute, or establish territorial claims and that they will assert no new claims while the treaty is in force. Under Article 10, the treaty States will discourage activities by any country in Antarctica that are contrary to the terms of the treaty. In view of these Treaty provisions, no new Entities below 60 degrees South will be added to the DXCC List as long as the Treaty remains in force.

c) The Spratly Islands, due to the nature of conflicting claims, and without recognizing or refuting any claim, is recognized as a Special Entity. Operations from this area will be accepted with the necessary permissions issued by an occupying Entity. Operations without such permissions, such as with a self-assigned (e.g., 1S) callsign, will not be recognized for DXCC credit.

d) Control of Western Sahara (S0) is currently at issue between Morocco and the indigenous population. The UN has stationed a peacekeeping force there. Until the sovereignty issue is settled, only operations licensed by the RASD shall count for DXCC purposes.

e) Entities on the 1998 DXCC List that do not qualify under the current criteria remain as long as they retain the status under which they were originally added. A change in that status will result in a review in accordance with Rule 5 of this Section.

4. Ineligible Areas:

a) Areas having the following characteristics are not eligible for inclusion on the DXCC List, and are considered as part of the host Entity for DXCC purposes:

i) Any extraterritorial legal Entity of any nature including, but not limited to, embassies, consulates, monuments, offices of the United Nations agencies or related organizations, other inter-governmental organizations or diplomatic missions;

ii) Any area with limited sovereignty or ceremonial status, such as monuments, indigenous areas, reservations, and homelands.

iii) Any area classified as a Demilitarized Zone, Neutral Zone or Buffer Zone.

b) Any area which is unclaimed or not owned by any recognized government is not eligible for inclusion on the DXCC List and will not count for DXCC purposes.

5. Removal Criteria:

a) An Entity may be removed from the List if it no longer satisfies the criteria under which it was added. However, if the Entity continues to meet one or more currently existing rules, it will remain on the List.

b) An Entity may be removed from the List if it was added to the List:

i) Based on a factual error (Examples of factual errors include inaccurate measurements, or observations from incomplete, inaccurate or outdated charts or maps); and

ii) The error was made less than five years earlier than its proposed removal date.

c) A change in the DXCC Criteria shall not affect the status of any Entity on the DXCC List at the time of the change. In other words, criteria changes will not be applied retroactively to Entities on the List.

SECTION III. ACCREDITATION CRITERIA

1. Each nation of the world manages its telecommunications affairs differently. Therefore, a rigid, universal accreditation procedure cannot be applied in all situations. During more than 50 years of DXCC administration, basic standards have evolved in establishing the legitimacy of an operation.

It is the purpose of this section to establish guidelines that will assure that DXCC credit is given only for contacts with operations that are conducted with proper licensing and have established a legitimate physical presence within the Entity to be credited. Any operation that satisfies these conditions (in addition to the applicable elements of SECTION I., Rules 6, 7, 8, and 9) will be accredited. It is the intent of the DXCC administration to be guided by the actions of sovereign nations when considering the accreditation of amateur radio operation within their jurisdiction. DXCC will be reasonably flexible in reviewing licensing documentation. Conversely, findings by a host government indicating non-compliance with their amateur radio regulations may cause denial or revocation of accreditation.

2. The following points should be of particular interest to those seeking accreditation for a DX operation:

a) The vast majority of operations are accredited routinely without a requirement for the submission of authenticating documentation. However, all such documents should be retained by the operator in the unlikely event of a protest.

b) In countries where Amateur Radio operation has not been permitted or has been suspended or where some reluctance to authorize amateur stations has been noted, authenticating documents may be required before accrediting an operation.

c) Special permission may be required from a governmental agency or private party before entering certain DXCC Entities for the purpose of conducting amateur radio operations even though the Entity is part of a country with no amateur radio restrictions. Examples of such Entities are Desecheo I. (KP5); Palmyra I. (KH5), and Glorioso Islands (FR/G).

3. For those cases where supporting documentation is required, the following can be used as a guide to identify those documents necessary for accreditation.

a) Photocopy of license or operating authorization.

b) Photocopy of passport entry and exit stamps.

c) For islands, a landing permit and a signed statement of the transporting ship's, boat's, or aircraft's captain, showing all pertinent data, such as date, place of landing, etc.

d) For locations where special permission is known to be required to gain access, evidence of this permission must be presented.

e) It is expected that all DXpeditions will observe any environmental rules promulgated by the administration under whose authority the operation takes place. In the event that no such rules are actually promulgated, the DXpedition should leave the DXpedition site as they found it.

4. These accreditation requirements are intended to preserve the integrity of the DXCC program and to ensure that the program does not encourage amateurs to "bend the rules" in their enthusiasm, possibly jeopardizing the future development of Amateur Radio. Every effort will be made to apply these criteria uniformly and to make a determination consistent with these objectives.

5. The presentation in any public forum of logs or other representations of station operation showing details of station activity or other information from which all essential QSO elements (time, date, band, mode and callsign) for individual contacts can be derived creates a question as to the integrity of the claimed QSOs with that station during the period encompassed by the log. Presentation of such information in any public forum by the station operator, operators or associated parties is not allowed and may be considered sufficient reason to deny ARRL award credit for contacts with any station for which such presentations have been made. Persistent violation of this provision may result in disqualification from the DXCC program.

SECTION IV. FIELD CHECKING OF QSL CARDS

QSL cards for new DXCC awards and endorsements may be checked by a DXCC Card Checker. This program applies to any DXCC award for an individual or station, except those specifically excluded.

1) Entities Eligible for Field Checking:

a) All cards dated ten years or less from the calendar year of the application may be checked, except for those awards specifically excluded from the program. 160 meter cards are currently excluded from this program. QSLs for years more than ten calendar years from the application date must be submitted directly to ARRL Headquarters. All deleted entities must be submitted to ARRL HQ.

b) The ARRL Awards Committee determines which entities are eligible for Field Checking.

2) DXCC Card Checkers:

a) Nominations for Card Checkers may be made by:

i) The Section Manager of the section in which the prospective checker resides.

ii) An ARRL affiliated DX specialty club with at least 25 members who are DXCC members, and which has DX as its primary interest. If there are any questions regarding the validity of a DX club, the issue shall be determined by the division director where the DX club is located. A person does not have to be a member of a nominating DX club.

iii) by Division Director.

b) Appointments are limited to one per section, one per DX club, and one per Division.

c) Qualifications:

i) Those nominated as a card checker must be of known integrity, and must be personally known to the person nominating them for appointment.

ii) Candidates must be ARRL members who hold a DXCC award endorsed for at least 150 entities.

iii) Candidates must complete an open book test about DXCC rules concerning QSL cards and the Card Checker training guide.

iv) The applicant must be willing to serve at reasonable times and places, including at least one state or Division ARRL Convention each year.

v) The applicant must have e-mail and Internet capabilities, and must maintain current e-mail address with DXCC Desk.

d) Approval:

Applications for DXCC Card Checkers are approved by the Director of the ARRL Division in which they reside and are appointed by the Membership Services Manager.

e) Appointments are made for a two year period. Retention of appointees is determined by performance as determined by the DXCC Desk.

3) Card Checking Process:

a) Only eligible cards can be checked by DXCC Card Checkers. An application for a new award shall contain a minimum of 100 QSL confirmations from the list and shall not contain any QSLs that are not eligible for this program. Additional cards may not be sent to HQ with field checked applications. The application may contain any number of cards, subject to eligibility requirements and fees as determined by Section I, Basic, 15.

b) It is the applicant's responsibility to get cards to and from the DXCC Card Checker.

c) Checkers may, at their own discretion, handle members' cards by mail.

d) The ARRL is not responsible for cards handled by DXCC Card Checkers and will not honor any claims.

e) The QSL cards may be checked by one DXCC Card Checker.

f) The applicant and DXCC Card Checker must sign the application form. (See Section I no. 11 regarding altered, forged or otherwise invalid confirmations.)

g) The applicant shall provide a stamped no. 10 envelope (business size) addressed to ARRL HQ to the DXCC Card Checker. The applicant shall also provide the applicable fees (check or money order payable to ARRL-no cash, credit card number and expiration date is also acceptable).

h) The DXCC Card Checker will forward completed applications and appropriate fee(s) to ARRL HQ.

4) ARRL HQ involvement in the card checking process:

a) ARRL HQ staff will receive field-checked applications, enter application data into DXCC records and issue DXCC credits and awards as appropriate.

b) ARRL HQ staff will perform random audits of applications. Applicants or members may be requested to forward cards to HQ for checking before or after credit is issued.

c) The applicant and the DXCC Card Checker will be advised of any errors or discrepancies encountered by ARRL staff.

d) ARRL HQ staff provides instructions and guidelines to DXCC Card Checkers.

5) Applicants and DXCC members may send cards to ARRL Headquarters at any time for review or recheck if the individual feels that an incorrect determination has been made.

Index

DXCC Item Order Form

The following are available from the ARRL DXCC Desk. Upon qualification, these items may be purchased. Prices subject to change without prior notice.

PLAQUES (Includes one engraved plate) (Allow 6-8 weeks for delivery)

- ☐ Top of Honor Roll ** ($55.00 plus Shipping)
- ☐ Honor Roll ** ($40.00 plus Shipping)
- ☐ 5 Band DXCC ** ($40.00 plus Shipping)
- ☐ Challenge ** ($79.00 plus Shipping)

REPLACEMENT PLATES FOR PLAQUES ($12 plus shipping)

- ☐ 5 Band DXCC ☐ #1 Honor Roll ☐ Honor Roll

ENDORSEMENT PLATES FOR 5 BAND PLAQUES ($8.50 plus shipping)

- ☐ 5 Band DXCC Endorsement Plate(s) (Must have 5 band plaque) Band(s): _____

LAPEL PINS (6.00 Each plus shipping)

☐ 2 Meter	☐ 6 Meter	☐ 10 Meter	☐ 12 Meter	☐ 15 Meter
☐ 17 Meter	☐ 20 Meter	☐ 30 Meter	☐ 40 Meter	☐ 80 Meter
☐ 160 Meter	☐ Mixed	☐ Phone	☐ CW	☐ Satellite
☐ RTTY	☐ 5BDXCC	☐ #1 Honor Roll	☐ Honor Roll	

REPLACEMENT CERTIFICATES ($12.00 Each plus shipping)

☐ 2 Meter	☐ 6 Meter	☐ 10 Meter	☐ 12 Meter	☐ 15 Meter
☐ 17 Meter	☐ 20 Meter	☐ 30 Meter	☐ 40 Meter	☐ 80 Meter
☐ 160 Meter	☐ Mixed	☐ Phone	☐ CW	☐ Satellite
☐ RTTY	☐ 5BDXCC	☐ Millennium	☐ QRP	

DXCC CHALLENGE PINS ($9.50 plus shipping)

- ☐ 1000 Level Challenge ☐ 1500 Level Challenge
- ☐ 2000 Level Challenge ☐ 2500 Level Challenge

DXCC CHALLENGE MEDALLIONS ($9.50 plus shipping)

- ☐ 1000 Level Challenge ☐ 1500 Level Challenge
- ☐ 2000 Level Challenge ☐ 2500 Level Challenge
- ☐ **3000 Level DXCC CHALLENGE MEDALLIONS ($15 plus shipping)**

☐ **THE ARRL DXCC List** ($5.95 each plus shipping)
☐ **ARRL DXCC Yearbook** ($5.00 each plus shipping)

Name: _____ Call Sign: _____

Address: _____

_____

Email Address: _____

Name as it appears on credit card: _____

☐ Master Card (Including Euro Card) ☐ VISA ☐ AMEX ☐ Discover ☐ Other: _____

Card # _____ Exp Date: _____

Fees subject to change without notice. Make checks payable to ARRL. All prices are pre-paid. ARRL not responsible for cash. Unless otherwise specified, we ship Fedex in the US, 1st Class to Canada and Air Mail Letter Post elsewhere. ** ARRL is not responsible for plaques lost in shipment that are not sent via a traceable method (such as Registered Mail, Fedex, DHL, or UPS).

ARRL *The national association for* **AMATEUR RADIO**

Please use this form to give us your comments on this book and what you'd like to see in future editions, or e-mail us at **pubsfdbk@arrl.org** (publications feedback). If you use e-mail, please include your name, call, e-mail address and the book title, edition and printing in the body of your message.
Also indicate whether or not you are an ARRL member.

Where did you purchase this book?
   ☐ From ARRL directly   ☐ From an ARRL dealer

Is there a dealer who carries ARRL publications within:
   ☐ 5 miles   ☐ 15 miles   ☐ 30 miles   of your location?   ☐ Not sure.

License class:
   ☐ Novice ☐ Technician ☐ Technician Plus ☐ General ☐ Advanced ☐ Extra

Name _____ ARRL member? ☐ Yes ☐ No
_____ Call Sign _____
Daytime Phone (   ) _____ Age _____
Address _____
City, State/Province, ZIP/Postal Code _____
e-mail address_____

If licensed, how long? _____
Other hobbies _____
Occupation _____

For ARRL use only	DXCC HBK
Edition	1 2 3 4 5 6 7 8 9 10
Printing	1 2 3 4 5 6 7 8 9 10

From _____

_____

_____

EDITOR, DXCC HANDBOOK
AMERICAN RADIO RELAY LEAGUE
225 MAIN STREET
NEWINGTON CT 06111-1494

———————————— please fold and tape ————————————